Careers in the Oil & Gas Industry: A Guidebook of Practical Advice

Alfonso Colombano

Ryan Ray

Copyright © 2018 Alfonso Colombano & Ryan Ray.

All rights reserved. No part of this book may be reproduced, stored in a retrieval system or transcribed in any form or by any means, electronic or mechanical, including photocopying and recording, without the prior written permission of the authors.

Book idea, compilation and authorship by Alfonso Colombano and Ryan Ray.

Self-published by Alfonso Colombano and Ryan Ray

Edited by Josh M. Shelton, Alfonso & Alberto Colombano and Ryan Ray

All company names, brands and slogans are property of their respective owners and are used throughout the book for reference purposes only.

ISBN-13: 978-1720733515

ISBN-10: 1720733511

Printed by CreateSpace, An Amazon.com Company

www.oilgascareers.net

Disclaimers

All data discussed and used in this book has been referenced and sourced from **publicly** available materials, such as companies' forms 10-K and 20-F, annual reports, internet websites, news articles, company reports, job postings and news releases. Reader should verify accuracy by checking cited references.

The recommendations, advice, analysis, descriptions, methods and calculations presented in this book are for educational and illustration purposes only. The authors shall not be liable for any income, career or any other loss or any damage that results from the use of any of the material in this book.

Alfonso Colombano, Ryan Ray or the book's publishers shall not be liable for any damages whatsoever, and in particular the authors shall not be liable for any special, indirect, consequential, or incidental damages, or damages for lost profits, loss of revenue, or loss of use arising out of or related to this book or the information within, whether such damages arise in contract, negligence, tort, under statute, in equity, at law, or otherwise, even if the authors have been advised of the possibility of such damages.

Table of Contents

Contents 5
About Alfonso Colombano 9
About Ryan Ray 11
Acknowledgements – Alfonso Colombano 13
Acknowledgements – Ryan Ray 15
Preface 17
Chapter I – Introduction 19
    Audience 20
    Why this book? 20
    Why work in Oil & Gas? 21
    Key Benefits of Reading this Book 22
    What is Upstream, Midstream and Downstream? 22
    Functions in the Oil & Gas Industry 24
    Functions or disciplines covered in this book 27
Chapter II – Types of Companies 31
    Introduction 32
    National Oil Companies (NOCs) 33
    International Oil Companies (IOC) 41
    Independent Downstream or Pure Play Refining Companies 49
    Independent Exploration & Production Companies 55
    Midstream Companies – Master Limited Partnerships (MLP) 66
    Service Companies 70
Chapter III – Talent Management & Career Development 75
    Historical Career Developmental 76
    Shell Oil and HAIR system 76

Traditional Oil Company model 77

Specialist vs. Generalist 78

Cost center vs. Profit Center 82

Curse of Competence 84

Ability to adapt 87

Breadth vs. Depth 89

Absolute vs. Comparative Advantage 90

Individual Contributor vs. Supervisory Assignments 91

Vertical vs. Horizontal Moves 92

Communication Skills 94

Nine-by-Nine Career Matrix 96

Pareto's Law 96

High Potential (Hi-Po) 97

Cross-Functional Assignments 100

Stretch Assignments 101

The Importance of having patience 103

Performance Management 106

Career Development Plans 109

Fast Lane Developmental Programs 110

Best practices in Talent Management 111

Public Speaking Skills 112

Chapter IV – The War for Talent in the Oil & Gas Industry 115

The importance of choosing the right career 116

Maslow's Hierarchy of Needs 117

Why work for an Oil & Gas company? 120

How to search for a job in the first place 120

Job Search Tips 125

Application Process Tips 126

Networking 126

Interview Process 127
Sample list of questions for a job interview 130

Chapter V – Accounting & Finance 133
What is Accounting? 135
What roles do accounting & finance play in the oil & gas industry? 135
Career Background 136
Why you should be in Accounting or Finance 137
Why you should not be in Accounting or Finance 137
Career Development in Accounting & Finance 138
Examples of Careers in Finance 142

Chapter VI – Engineering 155
What is Engineering? 156
Engineering Disciplines 156
Career Development in Engineering 157
Chemical Engineering 162
Electrical Engineering 166
Mechanical Engineering 169
Petroleum Engineering 171

Chapter VII – Other Career Groups 177
Operations 178
Geosciences 182
Information Technology 184
Business Development & Commercial 186

Chapter VIII – Conclusion 195
Summary 196
Assessing your current & future career state 196
Questions to ask when selecting a company to work for 197
Behaviors to observe when selecting a company to work for 197
Questions to ask a particular department 198

The Importance of an Action Plan 199
Action Plan for Early Career 199
Action Plan for Mid-Level & Late Career 200
Last Words 201
Glossary 202

About Alfonso Colombano

Alfonso is the author of the bestseller series of books titled *Oil & Gas Company Analysis*, which have been used as a teaching tool in the oil & gas industry worldwide. Alfonso has worked for two International Oil Companies (IOC's) and one refining company in analytical, commercial & financial roles. Alfonso is a graduate of the University of Houston where he was active in the Energy Association. Alfonso's expertise encompasses all sectors of the oil & gas industry including upstream, midstream and downstream, which led him to create, write and publish three books on the topic. Alfonso is an avid reader of Forms 10-K and 20-F, financial statements and other external company publications.

Alfonso's expertise encompasses the financial & operational analysis of oil & gas companies, having developed several analytical models for M&A activity and commercial analysis. Alfonso is a well-seasoned public speaker and has been featured at the University of Houston and the *Global Energy Leaders Podcast*. Alfonso is co-host with Ryan Ray on the *Energy Market Recap Podcast*. He enjoys reading books about the oil & gas industry and playing golf in his free time.

Alfonso can be reached by email at alfonso@oilgascompanyanalysis.com or by following his LinkedIn page at www.linkedin.com/in/alfonsocolombano

About Ryan Ray

Ryan Ray has worked in the energy industry since 2005. During that time he has served as the Director of Operations for R-Squared Global, given presentations on the on the oil and gas industry in the US and South Africa, published articles in two global trade magazines, hosted several energy-based podcasts such as The Global Energy Leaders Podcast, Oil and Gas Market Recap, Texas Oil and Gas Podcast, and Energy Week Podcast and, at the time of the writing, is a finalist for the "One to Watch" at the PEX Awards 2018

Ryan can be reached by email at ryan@globalenergymedia.com or by following his LinkedIn page at www.linkedin.com/in/ryanraysr

Acknowledgements – Alfonso Colombano

"Gratitude is a mark of a noble soul and a refined character. We like to be around those who are grateful." – Joseph B. Wirthlin

Without the immeasurable aid and support of my wife Indira, my son Massimo, my parents and brother who encouraged and assisted me, I would never have completed this lengthy effort.

Additionally, I express my gratitude to my co-author, co-host and good friend Ryan Ray who helped me immensely. I want to also express my gratitude to my good friends at the University of Houston, Bauer College and the Energy Association for having broadened my vision about the industry and having invited high caliber presenters from Oil, Gas and Energy related firms throughout my studies at the University of Houston.

Last, but not least, my sincerest thanks to all the great colleagues, supervisors, managers, operations personnel, and mentors that I have had the good fortune to have worked with in the energy industry throughout my career. In particular, I would like to express special thanks to Kathryn Epperson, for her great help, support and mentorship after all these years. These great colleagues and mentors encouraged me from the beginning to learn more about our industry, the interconnectedness of the different sectors, the importance of focusing on fundamentals and the potential of the industry to grow long-term. I have been quite blessed to have been surrounded throughout my career, including at university and at work, with brilliant people who have influenced my views and have helped me tremendously.

Without the help of all of you, I could not have completed this monumental task. Thanks everybody!

Alfonso Colombano
Houston, Texas, USA

June 2018

Acknowledgements – Ryan Ray

> *"He is a wise man who does not grieve for the things which he has not, but rejoices for those which he has."* – Epictetus

My wife is my best friend, closest confidant, biggest fan, harshest critic, and the hardest working person I know. She inspires me on a daily basis and makes me a better person. Thank you, Haylee, for all that you do.

I also need to thank my father, Rodney Ray, for giving me a job in the oil and gas industry in 2005. Without him it is extremely unlikely that I would be in the industry, much less writing a book about it. During my time in the industry, I have worked with a lot of great people. I would be remiss if I didn't mention my current and former colleagues at R-Squared Global. Thank you for all that you do on a daily basis.

Finally, I have had the pleasure of meeting and working alongside some great people who have helped me advance my career since breaking into the industry. Mike White, Clint Gregg, Muzi Shange, Josh Shelton, Kristen Underwood, Ellen Wald, Mark LaCour, and Alfonso Colombano who did the Lion's Share of the writing for this book.

Ryan Ray
Fort Worth, Texas, USA

June 2018

Preface

"Choose a job you love, and you will never have to work a day in your life." – Confucius

Dear reader,

Thank you for choosing this book. This is the fourth book I have published and I'm very excited to share with readers how exciting and fulfilling careers in the oil & gas industry can be. Many people, from all walks of life, background, formal education and locations, have enjoyed having a career or playing a part in *exploring, producing, transforming* and *marketing* hydrocarbons in this great industry. The Oil & Gas industry is a fascinating place to have a great, long-lasting and enjoyable career. Despite all the recent economic challenges in the industry, with commodity prices at recent historical lows, this industry is so vital to the world's economy that people should not think twice about starting a career here. More importantly, as the coming age demographic shifts[1] starts to impact succession planning, being ready and willing to advance your career becomes even more *critical*.

One of the goals of this book to provide readers with a detailed coverage of careers that are involved in exploring, producing, refining and marketing petroleum and petroleum products, from a diverse set of functions, such as Accounting, Business Development, Engineering, Geo-sciences, Operations, Marketing & Trading as well as support groups such as HR, IT and many others.

This book aims to answer the following questions:

- How to choose a career from the different functions available.
- How to choose the best type of company that fits your individual preferences, work-life balance ethic, career aspirations and many other factors.

---

[1] The demographic shift refers to the fact that last major hiring wave was in the late 1970's and early 1980's just before oil prices decreased significantly. There was very little hiring in the 1990's and early 2000's in the industry, creating a significant opportunity for new entrants.

- Provide *guidance* on what each major function does and pathways to entry into these careers or companies.
- Understand how to progress your career, how talent management works, research companies and apply for jobs within your existing company or another company.
- Understand what other functions in a company do, whether you work in a small company or a major integrated company.

The book is organized into eight chapters:

- Chapter I provides an overview of the oil & gas value chain, the different sectors and functions involved in this industry.
- Chapter II provides an overview of the different types of companies involved in the oil & gas industry, from large integrated oil companies to the smaller independent to everything in between.
- Chapter III discusses different concepts around *Talent Management* and *Career Development*, which are applicable not only to the oil & gas industry but to any business in general.
- Chapter IV provides advice on how to get a position with an oil & gas company, understand how to apply, understand the different types of interviews and sample list of questions to generate
- Chapters V through VII review several key functions involved in the industry, career paths, career progression for each function, guide to entrance and other commentary.
- Chapter VIII concludes the book and provides a call to action depending on your different career aspirations and current state.

We hope you enjoy this book and learn more about the exciting world of oil & gas careers.

Again, thank you for selecting this book!

Alfonso Colombano
Houston, Texas, USA
June 2018

Chapter I – Introduction

"Find out what you like doing best and get someone to pay you for doing it"
Katherine Whitehorn

The Oil & Gas industry is a fascinating business to be in. Just think about how often we are positively impacted every day by the products we use that are derived from hydrocarbons. From commuting to work, taking a trip from one corner of the world to the other, to life saving plastics, the oil & gas industry enriches and improves our lives. To produce a barrel of oil, transport it to a refinery and refine it into gasoline, requires the immense efforts of several diverse people from different backgrounds around the world. The role of this book is to provide background on each of the functions and career paths involved in the industry as well as to provide practical career advice on how to enter the industry.

Audience

Many readers can benefit from reading this book, particularly those new to the industry or looking to switch careers from other industries. Among some of the audience members who can benefit from reading this book:

- University students interested in careers in the different fields in the industry, whether they are pursuing a degree in engineering, accounting, geology, business or many others.
- Employees, contractors or suppliers working *in* or *outside* the oil & gas industry looking to broaden their understanding of the available careers and options in oil & gas.
- Experienced employees in the oil & gas industry who may work in one or function of a company but are interested in learning what other functions or groups do.
- Anybody interested in expanding their general knowledge about careers in this industry.

Why this book?

Choosing a career is one of the most *challenging* and *impactful* decisions one can ever make in life. We would venture to say that after choosing a person to get married to, choosing the right career is one of life's major decisions and can have *long-term impacts* in the quality of life of an individual. This book seeks to anchor *expectations* regarding career progression for the different paths available.

Different functions are involved in the oil & gas industry such as Accounting, Business Development, Engineering, Geo-sciences, Operations, Marketing & Trading as well as support groups such as HR, IT and many others.

The goal of this book is:

- Describe the typical functions and careers available in the industry.
- Guide people in choosing a career in this industry and describe the opportunities, challenges and typical activities each career path will normally be involved with.
- Detail the different type of companies in the upstream, midstream and downstream areas of oil & gas as well as the *pros* and *cons* of working for a *large* company vs. a *small* company.
- Provide guidance on how *talent management* is performed in the industry and how it varies from the *sector* as well as the *size* of the company involved.
- Describe the pros and cons of *specialization* (SME) vs. *generalization* as well the concept of *individual contributor* vs. *management* ladders.
- Provide examples of career paths, opportunities available, and challenges ahead in this industry.
- Provide pros and cons of each reviewed career path, including factors such as *work and life* balance, career portability, mobility and others.

Why work in Oil & Gas?

There are many reasons to work in the oil & gas industry and have a long fulfilling career, among which are:

- Expansiveness of operations, with *production* of oil & gas is located in many areas around the world, while *distribution* and *retailing* of petroleum products happening in practically every single country in the world.
- This industry is a highly capital intensive industry where, *compensation costs* as a percentage of *total costs* tend to be *lower* than other industries, such as services, retail and sales, therefore workers have a higher *relative* compensation than other industries.
- Specialization and complexity of the industry, which can lead to specialization and years or even decades of learning in a highly rewarding area.
- Going home realizing how impactful hydrocarbon operations are in the global economy and helping the world economy fuel growth.

- Highly segmented industry with several options to work for both an *operating company*[2] as well as *suppliers*.
- Extended career development options for all functions involved.

Key Benefits of Reading this Book

- Help readers narrow down their career choices within the multitude of options and describe the typical functions or careers involved in the oil & gas industry.
- Understand the different career paths available; better understand types of companies and philosophies of career & talent management.
- Develop an understanding of what other functions in an oil & gas company *do* on every day.
- Provide examples of when is better to specialize vs. being a generalist.
- Understand the different trade-offs between career stability, risk, visibility and career growth.

What is Upstream, Midstream and Downstream?

The oil & gas industry is traditionally divided into three distinct sectors, upstream, midstream and downstream:

- Upstream, also known as Exploration & Production, is primarily focused on finding, exploring, developing, producing and transporting raw crude oil and natural gas from the producing well to a terminal, gas plant, refinery or other delivery points.
- Midstream is one of the most difficult sectors to classify since the definitions of what constitutes "midstream" can vary from company to company. Conventionally, midstream has encompassed gathering, treating, processing of natural gas, long

---

[2] For the purposes of this book, "operating company" is defined as a company having or operating assets, while suppliers defined as a contractor company that supplies good or services. An example of an operating company could be Chevron while a supplier could be Halliburton or Schlumberger.

haul transportation of oil, natural gas, natural gas liquids, refined products and chemicals, as well as terminal assets, LNG, LPG and crude oil shipping, fractionation or separation of NGLs and other areas. For several companies, especially the large integrated companies, their "midstream assets" are usually grouped or reported as Downstream instead of as a standalone midstream segment.
- Downstream is principally involved in refining, marketing, transporting, delivering crude oil and natural gas and other products to wholesale and end customers. The downstream sector in the oil & gas industry also encompasses petrochemicals, lubricants, and other products.

The following table provides a summary of the activities, business models, price risk, on-going capex requirements, typical levels of profitability as well as other risks that are found in these three oil & gas sectors[3]:

	Upstream	Midstream	Downstream
Main Activity	Finding, exploring and producing hydrocarbons	Transporting and performing intermediate processing of hydrocarbons	Refining and delivering petroleum and petrochemical products
Business Model	Mining	"Toll-Road"	Manufacturing/Retail
Price Risk	Dependent on high commodity prices	Relatively low price risk. Primarily Fee-based earnings	Margin business, benefits from a high spread between *inputs* (crude oil) and *outputs* (petroleum products)
On-going CAPEX	High level, both on-going and initial capex	Initial capex high, moderate to low on-going capex	Initial capex high, moderate to low on-going capex
Profit Margin	High to Medium	Medium to Low	Low to Medium
Cash Flow Stability	Cyclical, relatively high	Most stable	Cyclical, relatively medium to low
Other Risks	Geopolitical & Environmental	Environmental	Environmental

---

[3] From Oil & Gas Company Analysis: Upstream, Midstream & Downstream by Alfonso Colombano

Functions in the Oil & Gas Industry

To find, explore, produce, process and deliver hydrocarbons, the work of extensive groups of diverse men and women, is required on a day-to-day basis from a multitude of functions.

Let's start by discussing the different areas or functions required in the upstream, midstream and downstream sectors at a high level.

Upstream (High level)

Beginning with the *Upstream side* of the business at a *very simplified* high level:

- A Landman is needed to acquire the necessary *mineral rights*[4] from mineral rights owners to *explore* for hydrocarbons in a particular area.
- A geologist and reservoir engineer are needed to study the rock formations and assess that subsurface's potential for producing hydrocarbons as well as to estimate the future hydrocarbon reserves from that well or reservoir.
- A drilling and completions engineer, drilling operations crew, roustabout and many people are required to effectively *drill* and *complete* the well.
- A reservoir engineer completes an assessment of the actual proved and proved developed reserves, which allows the company to invest in these wells and measure the economic performance.
- A production engineer is needed to assess and direct the *best way* to produce or *drain* an existing hydrocarbon reservoir as well as the installation of all the needed production, gathering & treating facilities.
- An environment, health and safety (HSE) employee is needed to mitigate risks and ensure the safe drilling and production of hydrocarbons.
- An operator or production person is needed to conduct all field related activities at the well, such as measuring volumes, conducting maintenance on wells, monitoring production, *tank strapping* to measure volumes correctly, dispatching trucks to pick up the oil, inspecting many wells and other activities.

---

[4] The United States is one of the few countries where it is possible to have private mineral rights ownership. Landmen positions usually do not exist in another country besides the U.S.

- An accountant is needed to correctly record and disburse revenues to working interest and royalty owners as well submit production and severance reports and taxes to the various state and regulatory agencies.
- A finance person is needed to understand and report all production and *post-production* costs associated with producing a barrel of oil as well as perform variance analysis on revenues, costs, and earnings.

Midstream (High Level)

Now transitioning into the Midstream side of the business, we describe at a very high level the typical functions involved in *building* a pipeline and *transporting* crude oil through this pipe (not necessarily in order):

- A commercial or business development starts identifying a business need for transportation services and develops a business case or proposal on why there is a need to have a pipeline built from point A to point B. This group then solicits bids from customers or *shippers* that would require this transportation service. After customers have signed up with long-term contracts, pipeline construction may begin.
- Procurement and project groups would oversee all aspects of the construction contract, such as opening up for bid the different areas of the project, planning and scheduling for the project construction activities, and estimate costs.
- A right-of-way group is tasked with acquiring the pipeline *right of ways* along the proposed pipeline route in order to start construction.
- A land surveyor is tasked with defining the exact boundaries of *where* the construction crew is expected to being excavating as well as making sure that there are no existing pipelines or any other utilities in the way so that construction can proceed.
- A mechanical engineer designs the pipeline thickness, pipe material, pump stations, and other equipment needed.
- A civil engineer oversees the excavation and pipe laying process.
- Operations & construction personnel conduct the process of building the pipeline.
- A finance or project controls employee is tasked with overseeing all financial aspects of the project, such as paying vendors, ensuring

budgets and commitments are met as well as validate the ongoing economics of the project.
- A legal or compliance employee is assigned with filing *tariff rate cases* with either the Federal Energy Regulatory Commission or FERC[5] or the local regulatory agency.
- The pipeline begins operations, schedulers and pipeline control personnel are assigned so that volumes can be safely transported through the pipeline and each shipper is allocated their correct volumes.
- Health, Safety & Environment personnel provide compliance services to the pipeline, such as safety filings with the Department of Transportation and safety agencies and ensure the construction process goes smoothly without any safety or environment incidents.
- An accountant administers the tariff and bills each shipper for their cost of transportation, as well as records and reports all financial transactions, such as paying vendors, income tax filing, financial statement analysis, and other responsibilities.

Downstream (High Level)

Finally on the downstream side of the industry there are multiple businesses in this sector as well as usually being the most *end-customer-focused* sector. The following provides a *high level* overview of the functions involved in *processing* raw crude oil and delivering motor gasoline to a retail station:

- Chemical engineers, along with supply & trading operations personnel, run *Linear Programming*[6] models to determine the *best* and *most* economical type of crude oil that the refinery can run to maximize profitability.
- A crude buyer then goes out into the market and procures this type of crude oil at the lowest possible cost.
- A crude scheduler then performs all logistics responsibilities to get that crude oil into the refinery's inventory tanks as *safely* and *timely* as possible.
- Refinery operations personnel analyze the crude oil with crude assays[7], ensuring that the delivered crude oil meets the

---

[5] FERC, in the United States, is in charge with all *economic* regulation of pipelines that cross state borders.
[6] http://www.eudoxus.com/mp-in-action/oil-industry/mpac9703
[7] A crude oil assay is essentially the chemical evaluation of crude oil feedstocks by a testing laboratory. https://www.statoil.com/en/what-we-do/crude-oil-and-condensate-assays.html

requirements of the refinery and as well verify the quality that the crude seller claimed that crude oil had. If the crude oil is not up to spec, this batch of crude oil is rejected and the seller is penalized.
- Refinery operators schedule the *distillation* of this crude oil into its different hydrocarbon components or *petroleum fractions*[8] which are then further processed into the refinery.
- A chemical process engineer evaluates the optimal blend of catalysts[9], to maximize desired yields of key refined products, such as motor gasoline, diesel and jet fuel.
- A maintenance employee ensures all the process and rotating equipment in the refinery is maintained in the most *efficient* possible way and all *unplanned* or *corrective* maintenance is minimized.
- A refinery accountant ensures that charge & yield[10] process is conducted correctly and all inputs and outputs into the refinery are accounted for accurately and timely.
- A refined products petroleum marketer or trader places a certain number of barrels into wholesale marketers which then sell these products into service stations across a region.
- A refined products scheduler ensures that the output of refined products is placed into the desired pipelines or modes of transportation to reach the desired end customers.
- An independent petroleum marketer then buys these barrels and sells them at their service stations.
- A service station buys gasoline, places it in its tanks, and sells it to motorists that stop at the station.

The previous overviews, as described, are at a *very high level* and *simplified* to illustrate high level positions. The amount of functions and activities performed in the entire oil & gas industry is quite staggering.

Functions or disciplines covered in this book

As discussed in our very simple overview in the prior pages, many functions are involved in the oil & gas industry. While every attempt has been made to incorporate the highest number of functions or disciplines into this book

---

[8] Petroleum fractions are the distinct group of hydrocarbon components that make up a particular crude oil and that share similar properties, such as boiling points.
[9] A refinery catalyst allows a chemical reaction to take place faster, as it the case with platinum based catalysts used in the fluid catalytic cracking process
[10] Charge & Yield is a process whereby the refinery is *mass* and *energy* balanced so that all inputs and outputs are accounted for correctly as well as a price placed on each component.

the following functions are covered in this book in different levels of details:

Accounting & Finance

Accounting or Finance functions are involved in the *accurate* and *timely* accounting, processing of transactions and financial reporting of all activities in an oil & gas company. This function is covered in more detail in Chapter V.

Engineering

Engineers are an integral part of the oil & gas industry and participate in a wide variety of functions. This function is generally involved in the *economic* and *practical* application of science to solve everyday challenges in the production, transportation and processing of hydrocarbons. Engineers have different sub disciplines, such as chemical, electrical, mechanical or petroleum engineering. This function is covered in Chapter VI.

Other Career Groups

Other career groups are covered in Chapter VII of this book. Below is a short description of the different groups involved.

Business Development, Commercial Operations, & Trading

This is an all-encompassing function or discipline that is akin in other industries to *sales* or *marketing* functions. This function is often engaged in negotiating agreements, buying and selling of commodities such as oil, natural gas, natural gas liquids, liquefied natural gas, and secondary byproducts.

Geoscientists

Although primarily engaged in the upstream sector of the industry, the geoscience function plays an integral part in the *exploration* process of discovering and producing hydrocarbons.

Operations

Operations is a *catch-all* function for the variety of activities needed to produce, transport, process, and market hydrocarbons today. This function covers a wide variety of positions, such as refinery maintenance, scheduling/dispatching operations, pipeline operations, well operators, construction workers, electrical crews, and many others.

Information Technology

Information Technology plays a key role in all phases and functions within the oil & gas industry. As reliance on technology continues to increase and more and more functions are automated, a good understanding of how the IT function works as well as how key IT process might impact other functions is warranted.

Support Functions

Similar to operations, support functions are generally not cataloged in terms of a specific discipline, but it is a conglomerate of different functions. The difference here between "support" and "frontline operations" is that support functions are generally *not* considered *front office* or frontline, but commonly play more of a support or corporate role *"back office"*.

Chapter II – Types of Companies

*"All you need in this life is ignorance and confidence, and then success is sure"–
Mark Twain*

Introduction

The oil & gas industry, like similar industries, encompasses companies from all over the globe, from all sectors and with different types of ownership (state, private or publicly traded). Each of these types of companies has different career and talent development philosophies, different ways of conducting operations, risk tolerance approaches and compensation & benefits structures. In this chapter, the following companies are profiled:

- National Oil Companies (NOCs).
- Integrated Oil Companies or International Oil Companies (IOCs).
- Independent Downstream or Pure Play Refining Companies.
- Small, Medium, and Large Independent Exploration & Production Companies.
- Midstream Companies.
- Service Companies.

Each of these companies is presented and assessed with the following criteria:

- Career Progression or the speed at which a typical employee would progress through a career ladder.
- Career Stability, how volatile or stable a career would be a in a company taking into account the business cycle that the oil & gas industry experiences every few years.
- Culture, although each company's culture is unique, there are many traits and characteristics that are shared between companies within the same categories.
- Risk Tolerance, how *risk adverse* or *risk taker* each type of companies is and how risk perception impacts careers in a company.
- Risks Involved, what are some of the risks involved in working for a different type of company, such as *low visibility* risks in a larger company or *career stability* risks in a smaller company.
- Compensation & Benefits, although each company is unique in a sense, many companies within the same subsector will tend to have similar compensation & benefits philosophies.
- Performance Management, how companies assess and evaluate performance for employees.

National Oil Companies (NOCs)

National Oil Companies or NOC's are the biggest participants in the upstream oil & gas industry in terms of *reserves* and *production*. Beginning in the 1950's and continuing through the 1980's, the creation of NOC's became more prevalent. One of the main drivers for the creation of NOC's was that host governments wanted to increase participation in the oil & gas industry in their own countries. NOCs can be further divided into two categories:

- Non-Publicly Traded National Oil Companies
- Publicly Traded National Oil Companies

Non-Publicly Traded National Oil Companies

As their name implies, these companies are fully owned and/or controlled by their respective governments and do not allow individual investors to own shares in these companies. These companies are among the largest companies in the world and are key players in the oil & gas markets. They are usually integrated since they have upstream, midstream, and downstream operations. Examples of some of the companies in this category are PDVSA, PEMEX, and Qatar Petroleum.

Publicly Traded National Oil Companies

Publicly traded NOC's have a substantial ownership or control by their respective host governments, but allow individual investors to own shares in these companies. These publicly traded NOCs are usually listed on major stock indexes around the world such as New York Stock Exchange (NYSE), London Stock Exchange (LSE), Moscow Stock Exchange (MOEX), among others. These companies allow investors to buy and sell shares of their companies on exchanges around the world through what are known as American Depositary Receipts or ADRs[11] (also known as Global Depositary Receipts or GDRs). These companies may also invest beyond their origin countries and are expected to compete more and more in the future with the International Oil Companies (IOC). Examples of these companies are Statoil, Petrobras and Petrochina.

The list on the following pages provides an overview of the different aspects of working for an NOC.

---

[11] For more information, please visit http://www.investopedia.com/terms/a/adr.asp

Advantages

There are many advantages in working for an NOC:

- Career stability, with companies having *outright* ownership by governments or *explicit* or even *implicit* financial backing by their governments. The likelihood of an NOC going out of business is far *lower* than a *publicly traded* or private company. Another reason for career stability is that many NOC have employment clauses with their home country nationals that impose restrictions on layoffs or firing of employees.
- The concept of being an employee for life or a *lifer*, which in today's fast pacing and ever changing economy, is difficult to find in any company, let alone in the oil & gas business.
- Access to large training programs, with many of the NOCs, along with the IOCs, having the most comprehensive, all-encompassing training programs.
- Career progression through different segments and businesses, with many NOC's having operations in upstream, midstream or downstream.
- International exposure, with now several NOC's owning or operating assets in locations far beyond their home countries.

Disadvantages

There are several disadvantages in working for an NOC:

- Decisions made by *consensus*, involving several committees inside and outside the NOC, so these companies are generally not as agile as smaller private companies. National Oil Companies and their corresponding Ministry of Energy positions are always navigating a fine balance between oversight, governance, and micromanagement.
- Decisions sometimes are not taken from a purely economic point of view, but may involve politics and societal impacts. For example an NOC may decide to operate *loss-generating* service stations because they provide significant employment to country nationals.
- Politics could impact hiring and firing decisions, which may not work the most optimal for outsiders or those with not significant political connections.
- Employment might be reserved for local nationals and expat opportunities could be limited.

- Lack of detail focus and granularity that many private or even publicly traded companies could have. With the companies' size and extensiveness of operations, it becomes quite impossible to *know the pulse* of what's happening across the company. Many functions could operate in a *vacuum* or *silo*.

Future Opportunities

As National Oil Companies continue to become more international and their local human capital increases there will be diminished opportunities to become an expat[12]. One area of growth that we see in the future is to become a contractor for companies instead of being a *full-time employee*[13]. Another advantage of this strategy is the fact that there might be no visa, immigration, or quota restrictions imposed on contract labor, particularly those who work remotely outside of the NOC's home countries. Opportunities for locals in countries outside their original region will also increase. Recent examples of these employment growth opportunities in the US, where NOCs have opened local affiliates are Petrochina, Petrobras, Statoil, Aramco Services, Gazprom and many more:

- CNPC USA, a subsidiary of CNPC or Petrochina is headquartered in Houston and was founded in 2011[14].
- Petrobras USA, a subsidiary of Petrobras, has offices in Houston and New York, and has operations in Upstream, Downstream and Trading[15].
- Aramco Services, is the U.S. based subsidiary of Saudi Aramco and is headquartered in Houston[16]. This company is responsible for managing three U.S. research centers, identifying technologies and best practices in upstream and downstream, as well as recruiting professionals for careers in Saudi Arabia (expat opportunities).
- Statoil has extensive operations in the United States, where it operates or owns more than 2,200 onshore producing wells, 9 offshore fields, and produces about 250MBPD of oil equivalent[17].

---

[12] What is an expat? An expat or expatriate is an individual from one country, generally a developed country, who *temporarily* lives in another country. Expats are usually highly compensated in comparison to local population workforce since they bring specialized skills and higher productivity.
[13] A full time employee usually has more long-term benefits such as insurance, retirement plans and other, but often at a lower salary than a contractor. A contractor needs to pay all these expenses out of pocket and it is typically a short-term nature.
[14] http://www.cnpc-usa.com/about-us/corporate-profile/
[15] http://www.petrobras.com/en/countries/u-s-a/operations/
[16] http://www.aramcoservices.com/Who-We-Are.aspx
[17] https://www.statoil.com/en/where-we-are/united-states.html

The main offices for Statoil USA are located in Houston and the company currently employs more than 1,200 people[18].
- Gazprom USA, which is headquartered in Houston currently markets and trades natural gas throughout the U.S. as well as participates in foreign exchange trading. This subsidiary is also engaged in LNG trading and employs less than 200 people[19].
- EcoPetrol American Inc, is a U.S. subsidiary of Colombia's national oil company, has had operations in the US since 2007. This subsidiary is headquartered in Houston and has operations in the Gulf of Mexico in the upstream and employs less than 200 people[20].

Career Development

Most, if not all, NOC's provide structured development programs for their full-time employees. Often these are run in terms of function or location. Below are a few examples of career development and programs among representative NOC's:

Saudi Aramco

Saudi Aramco has one of the most comprehensive training programs in the industry, what is called Professional Development Program. New college graduates who have been sponsored by Saudi Aramco, or are hired directly from college, participate in a 3-year Professional Development Program (PDP). Throughout the PDP program, participants work with experienced employees to accelerate their development into fully qualified professionals. Participants requiring further training in English are enrolled in the Professional English Language Program (PELP), extending their stay in the program by up to 1 year. This program is used to identify those employees with the most potential for advancement to management positions[21].

Gazprom

Gazprom, the world's largest natural gas producer, has a comprehensive support program whereby they partner with key leading higher education institutions to provide interns and future employees of the company. In fact, Gazprom has two dedicated colleges owned or operated by the company which provide higher education, such as the Gazprom College

---

[18] Statoil 2016 Annual Report and Form 20-F, page 78
[19] http://www.gazprom-mt.com/WhatWeDo/Pages/Houston.aspx
[20] https://www.ecopetrol-america.com/index.php/who-we-are
[21] http://www.saudiaramco.com/en/home/careers/saudi-applicants/benefits/employee-training-programs.html

Volvograd, which enrolls more than 1,000 students per year and provides new candidates to Gazprom[22]. Competition for sought-after jobs in Gazprom is so high that the company organizes an annual competition among college students of different Russian universities, with winners getting awarded prizes as well as a guaranteed employment offer in Gazprom[23].

Kuwait Petroleum Corporation
Kuwait Petroleum Corporation or KPC provides extensive training and development programs for its employees. KPC has a dedicated location, called the Petroleum Training Centre, which can accommodate several employees at the same time in more than 29 training rooms, labs and computer rooms in Kuwait[24]. In addition, KPC has strong links with universities developing local talent by sponsoring education opportunities at foreign universities[25].

ENI
ENI, the Italian NOC, has always placed a high value in training, with one of the earlier training programs being created by the company's very own founder Enrico Mattei in 1957, with the founding of the School of Highest Studies in Hydrocarbons, now named "Scuola Mattei". Currently the company provides access to the Eni Corporate University, which is responsible for the selection, training and career development of the company's employees[26].

Career Stability
NOCs, due to their government backing, tend to be amongst the most stable companies from a career perspective. In fact, NOC often see employees as having more or less a *lifetime contract*. Saudi Aramco's expatriates are called "Aramcons", and have enjoyed lifetime contracts with the company as well as best in class benefit programs[27]. Another factor that enhances this lifetime employment contract culture is the fact that it is extremely difficult to get hired by these NOC's on the first place. Getting hired is so difficult that it generally requires having significant powerful connections, friends in high places, or high level networking. Typically once

---

[22] http://www.gazprom.com/careers/education/institutions/vcgo/
[23] http://www.gazprom.com/careers/education/competition-for-young-professionals/
[24] https://www.kpc.com.kw/OurPeople/Pages/Training-Centre.aspx
[25] https://www.kpc.com.kw/OurPeople/Pages/Developing-our-People.aspx
[26] https://www.eni.com/en_IT/careers/training-and-guidance/eni-corporate-university.page
[27] http://www.nytimes.com/2004/03/16/business/aramcons-find-arabia-like-home-sort-of.html

an employee gets hired by an NOC, the likelihood of finding another employment opportunity with a company of similar overall benefits is *low*.

Sometime this career stability may also cause problems in terms of spending for the overall company. Frequently most NOC's tend to be *overstaffed* in comparison to publicly traded or private companies since many of the positions available are seen as *national employment* programs.

Culture

Culture among NOCs can be quite different depending on the home country they are headquartered in as well as how the company is managed. Some NOCs still retain certain remnants of their historical company culture inherited from the International Oil Companies, such as performance management systems, career development, risk assessment, and many other areas. Such is the case with companies like Saudi Aramco (which as late as the 1980's had partial IOC ownership). This can be seen in many of these companies retaining such IOC cultural items as *expatriates*, living compounds, educational benefits, comprehensive training programs and sponsorship of advanced degrees. Over the past couple of years, NOC's have expanded their influence in the industry and increasingly rely on service operators to perform various types of work. With this change on relying on service companies, company cultures have changed and have become more in line with global company standards.

Risk Tolerance

In terms of risk tolerance it depends on the actual NOC and the governance. Some NOC's, like Petrobras or Statoil, are much less risk adverse than say a SaudiAramco or Gazprom. Petrobras in particular, because they had implicit backing by their government, grew and expanded very rapidly in offshore drilling knowing full well that if something were to happen, they would be *"rescued"* or *"bailed out"* by the Brazilian government. Petrobras being aware of this dynamic expanded very rapidly by issuing large amounts of debt and performing different expansions that turned out to be misallocated.

Recruiting

Getting into of the NOC's in general tends to be highly complex due to high demand and desirability of jobs in the oil & gas industry, not just for local employees but also to become an expat. Recruiting for expat opportunities is highly selective and in most cases requires ten plus years of prior industry experience before even being considered. Recruiting at local

NOC home countries varies on different factors, such as local examinations, networking or connections available and recruitment calendars. Often local NOC positions are among the most desirable jobs in these countries and competition will tend to be fierce. Many NOC's are also now implementing freezes on hiring foreign workers and we can reasonably conclude that the amount of expat positions in the future will be less than in the past. Companies like Saudi Aramco have even implemented systems where they track the nationality of workers for not only its direct employees but also for contractors' workforce, such as its well-known *Saudization* program[28].

Risks Involved

There are several risks involved in working for a National Oil Company that are worth mentioning:

- Similar to the discussion on IOC's, because of their size and number of employees, most NOC employees have a risk of *hyper specialization*. In NOC's, like in most large companies in the oil & gas industry, specialization and deep knowledge of a particular subject or range of subjects is highly sought after. If you are interested in becoming more of a generalist, these positions are usually not available within these companies to the majority of employees or will take many years to achieve.
- Slower career progression, particularly in comparison with independent private companies.
- Depending on the NOC, some companies have risks associated with political decision making or government relations. In NOC's with strong governance models and independence from their governments, this might not be the case. In many instances a change of government has brought about a change in the top management executives of an NOC but also as well as middle management. Usually, entry level, management and functional specialists or SME positions are often more stable than those that depend on political appointment.
- Slower decision making as well as heavy influence of politics, such as *local content requirements*, being that even if a particular contractor might not be qualified or there might be a better foreign contractor

---

[28] http://www.saudiaramco.com/content/dam/Publications/Saudization_Planning.pdf

available, the company would typically select the local contractor because of tough local content requirements.
- Weaker internal controls when compared to publicly traded companies, increasing the probability of mismanagement resources.

Compensation & Benefits

Compensation & benefits at NOCs tend to be the highest if not the highest of all companies in the oil & gas industry. In line with many NOC's philosophy of *lifetime employment,* benefit packages for these companies are highly competitive and look to retain talent for the entire career of that person. Among the benefit perks available to several NOCs[29]:

- Annual vacation benefits of up to 38 days plus national holiday days, anywhere from 9 to 11 days[30].
- Education Assistance plans, which covers dependent children primary and secondary education, including college, usually outside of the home NOC's country.
- Free or subsidized housing at company compounds or facilities, with spacious apartment or houses for the worker's nuclear family.
- Annual repatriation allowance to pay for airfare and other transport costs back to an expat's home country.
- Highly competitive bonus plan, typically in line or higher than most IOC's.
- Maternity and paternity leave plans that often exceed most private companies' benefits.
- Extensive retirement, savings and insurance plans, with a wide variety of domestic and international health, life, and other types of insurance.
- Most NOC's, due to the strong focus on long-term careers and lifetime employment provide top notch pension plans, which cannot commonly be found in private or publicly traded companies.

Performance Management

The performance management process is different in NOCs than other public or private oil & gas companies in many ways:
- Absence of externally published financial and valuation metrics.

---

[29] http://www.saudiaramco.com/en/home/careers/Non-Saudi-applicants/employment-benefits.html
[30] http://www.aramco.jobs/Comp-Benefits/Benefits.aspx

- Since the majority of these companies are not listed on stock markets, there is no concept of increasing shareholder value per se.
- Generally there is no *forced ranking* since the threat of layoffs or firings is lower than IOCs, with many employees having local union contracts that protect them from dismissals.

Most NOCs tend to focus on the following metrics to assess overall company performance:

- Safety metrics, such as total recordable injuries.
- Revenues generated or paid to the national government, set a budget price for the year.
- Hydrocarbon production targets or quotas set by the NOC or its corresponding Ministry of Energy.
- Individual department or group metrics, such as low percent of corrective maintenance, refinery utilization or uptime, percent of downtime or many other business-oriented performance management metrics.

In many other ways, the performance management process is similar to other large companies:

- Goals are cascaded centrally through the top levels of the company and are pushed down with more detail to lower levels and functions throughout the company.
- Employees' performance is assessed against similar compensation bands and experience to assess overall relative performance.
- Higher *relative* performance rankings would result in higher salary increases than the average.

International Oil Companies (IOC)

These are the companies most traditionally associated with the oil & gas industry such as ExxonMobil, Shell, BP and Total. These companies are *integrated* in the sense that they have operations in all three sectors, *upstream*, *midstream*, and *downstream* around the globe.

Career Progression

Career progression tends to be highly structured and moves in a more or less predictable manner. Career progression is usually handled by *functional talent management* teams or committees that provide oversight to an employee's career, beyond the employee's immediate supervisor or manager. These teams usually plan out ahead talent rotation and conduct succession planning so that companies always have a consistent pool of employees available to step up and fill these roles in case they need to be filled on a short notice.

Another consideration as to why companies may rapidly progress younger employees through the ranks faster than before is that as many more experienced employees retire, instead of backfilling these positions with outside hires, a cost containment measure is to promote internal younger employees, which are often lower cost than employees with more years of service.

Location is another important underpinning the operations of these global multinational corporations. Expatriate or *expat*[31] opportunities have been diminishing in comparison to the 1970's through the mid 2000's, but being open to relocation to any country can significantly increase the chances of moving up the corporate ladder faster. Expat opportunities are an excellent *stretch assignment*[32] since a person being transferred to a new location or country would be involved in a more *end-to-end* perspective than with a more *specialized* role in the corporate home office.

Last but not least, one of the key factors in career progression in a large integrated company is the point of entry and whether a person is considered *high potential*[33] or not. If an employee is labeled early on their career as a high potential employee, this employee would typically experience faster career progression than *non-high potential* employees. Please refer to the chapter on career development for a more in-discussion of high potential employees.

Career Stability

Career stability tends to be amongst the highest in the oil & gas industry, which is primarily due to their integrated business model, so when the upstream business does well, the whole company benefits and certain functions might be able to transfer from one business unit to the other. Another factor to consider is that career stability also depends as well on the function itself. For certain integrated companies, *back office* functions, like Finance, IT & Procurement, have been either *outsourced* or *offshored* to

---

[31] Expat or *expatriate* are usually personnel that is assigned to work in a different country from their home country. For example, a British employee of BP, who started working in London and is then transferred to Houston would be considered an expat.

[32] Stretch assignment is defined further book, but it is basically a work assignment that *stretches* or makes you grow in your career and capabilities.

[33] High potential is defined in more detail further along the book, but basically is defined as an employee that has the *potential* to be promoted several levels beyond their current position.

lower cost locations outside of the United States or their primary operating locations. For these functions, a career in a super major might be less *stable* than in a smaller company. Another aspect impacting career stability is the fact that the majority of positions in a company lack the overall *visibility* with senior management that another employee would have in a smaller company.

Culture

Culture is highly important in an integrated oil & gas company and most companies would look into having a typical *strong cultural fit* before hiring a new employee. Not only does culture vary around the type of company, but also can vary based on the countries the company operates in. Another impact into a company's culture is the corporate culture of the *heritage* companies prior to mergers and acquisitions.

Here are some the typical aspects of a large integrated oil & gas company:

- Places high level of emphasis on *communication* and *management of change*, due to a company's size and diverse groups and functions that are required to understand a change before it is implemented.
- Values *compliance* over *creativity*, particularly in those functions that operate in an environment with high levels of regulation (i.e. Securities & Exchange Commission financial reporting, Environment, Health & Safety compliance and HR).
- High emphasis on *standardization* and process focus versus understanding the *why* and the *end-to-end*.
- Career development, particularly for *high potential* employees, is highly valued and companies usually take risks and short-term inefficiencies in an effort of developing key employees and expanding their knowledge.
- Complex inter-business relationships and natural *silos* created among different geographical areas and business functions where the company operates.
- Strong emphasis on corporate systems across the company to achieve as *homogenous results* as possible.
- Highly *structured* decision hierarchy and *low level* of autonomy for most items at the *lower level*, which has both benefits and drawbacks.

- Large degree of influence from *cost center*[34] groups vs. *profit center* groups. This is generally the case because of the high emphasis on compliance, so a cost center group like Finance, IT or HR can exercise a significant influence on operations or engineer.
- For most groups and functions, there is an inherent *inability* to trace results to an individual or a few individuals, particularly applicable in the support groups, such as IT, Finance, Legal and others. In other words, how much of the success of the company this year is attributable to front-line operations vs. back office functions? A difficult question indeed.
- Slower decision requiring multiple teams' inputs and feedback, which can have both pros and cons. One pro being that results are less variable and more predictable.

Risk Tolerance

Integrated oil companies, because of their size, history, diverse global operations tend to be the most *risk adverse* companies in the industry. Although with the recent downturn in oil prices, many of the risk tolerance seen in the past is beginning to fade, the culture of *risk avoidance* at all costs still dominates the majority of integrated companies.

One of the many advantages of this extensive risk avoidance culture is the fact that these companies have the highest safety records of not just companies in the oil & gas industry, but across all industries. These companies usually have large *Environment, Health & Safety* departments (also known by acronyms such as HSE, HES, or EHS) that are highly qualified and can have a very positive influence in the company's operations. This is particularly important if the reader is interested in working in a high risk occupation such as construction, field operations or other. For example a mechanical engineer working out of college would have a more safety-focused culture by working at an IOC than by working at a small independent company, where corporate resources are less extensive. This is a key consideration when making a decision of which company to choose, the safety record.

"Safety does not happen by accident" - *Anonymous*

---

[34] The definition of cost center and profit center is further defined in the book. Cost center groups are usually that do not have traceable *profit & loss* indicators versus a profit center group which can have traceable P&L

Recruiting

Recruiting is often *highly structured* and managed *centrally* by the HR function. Recruiting tends to be more heavily weighted towards university recruiting, with examples like the U.S. having a busier recruiting season linked to the Fall Semester than the Spring Semester for major universities across the country. Integrated companies do hire for *mid-level* to *management level* positions, but the traditional path is usually centered on *lifetime employees*. These companies also have highly structured internship programs which are typically used for identifying *high potential* employees early-on in their careers and aim to create *long-term* loyalty between employee and employer. Many companies use summer internships as basically a *three-month interview* for full time candidates. These internship programs may grant participants direct access to senior and executive management that most employees would not have access on a day-to-day basis.

Because of the way recruiting is structured, the interview is highly programmed with questions commonly following techniques structured around STAR or Situation, Task, Action & Result or other *behavioral based* interview questioning systems.

Risks Involved

There are several risks or downsides involved in having a career in an integrated oil & gas company:

- Lack of visibility, especially earlier in your career. Being that these companies have billions of dollars of investments, several business units and thousands of employees, it is difficult to see how one employee's contribution might impact the organization's overall goals.
- *Hyper specialization* where due to the massive size of these companies, an employee could very well stay within a function or department for their entire career and still achieve a moderate to high compensation. For some people this may not be the most desirable career option.
  - Slower career progression, since there are more lateral positions available and far in between management or supervisory positions.
- Unnecessary risk avoidance, whether *real* and *manageable* risks or *perceived* risks, the integrated companies tend to avoid risks, sometimes at all costs. Again, similar to the discussion on safety, this has both positive and negative implications.

- Slower decision making, since decisions generally have to be made through consensus and any change by definition involves a large number of participants; these companies have to follow a structured process instead of the faster decision making found in smaller companies. Slower decision making has many benefits as well, such as the many IOC's which do not jump on the latest *fad* or bubble and require markets to develop more and be more stable before entering that particular market or region.
- Over-reliance on process and procedures, which for some personalities may not be the best fit. If you are an out of the box thinker and like to think around edges, these companies are built on standard processes to minimize variability, which may not work out.

Compensation & Benefits

Salaries in the integrated companies tend to generally be lower than those from both independent upstream and downstream companies. Salaries are usually grouped into "salary bands", "payscale groups" or "salary grade levels" with highly structured progression and "time in grade" requirements through each individual grade level. Because of the *consensus* based approach at making decisions, *relative performance* and salary adjustments are difficult to make on an individual basis and are assessed mostly through yearly assessment process.

Benefits in the integrated companies are higher, with many companies offering both a traditional 401k plan with generous matching as well as a pension plan. For newer employees these pension plans are usually *cash balance* plans instead of truly *defined benefit plans*, but are amongst the vest pension plans of any industries. Another area where integrated oil companies tend to exceed is in time-off from work, with companies having some of the best time-off policies of any companies, extensive vacations, compressed schedules such as 9/80[35], maternity and paternity leave, medical leave or personal leave. The oil & gas majors are one of the few company groups that still provide retirement benefits, and not only that but within their retirement plans have a lower percent of capitalization versus other companies. In a recent survey these companies had pension benefit obligations of about 22%, significantly lower than companies such as GM,

---

[35] A 9/80 schedule is where workers the same 80 hours of work in 2 weeks but instead of working 10 days they work more time in 9 days, therefore getting generally every other Friday off from work.

Ford, and other major companies with traditional pension plans[36]. Moreover in the same recent survey, integrated oil & gas companies spend the most in terms of percent salary of retirement benefits, with an average for all majors of 13% of pay, compared with less than 5% for manufacturing[37].

Many integrated oil & gas also sponsor educational reimbursement programs that go far and beyond the typical IRS-limited reimbursement programs. Company sponsored MBA's and other advanced degree programs are usually reserved for *high potential*[38] employees which the company sees as having management potential. In fact, whenever a company desires to sponsor an employee to achieve a higher degree this is usually confirmation that this employee is in the exclusive high potential club[39].

Sample Benefits from ExxonMobil

The paragraphs below examine ExxonMobil's current employee benefits for the year 2018 for U.S. based employees.

- Savings or 401K matching, with employees contributing 6% of pay, including salary and bonus, with the company matching 7% of the employee's pay to a 401K account on a before-tax basis[40]. The 401K savings is originally matched in ExxonMobil stock, but participants can choose from a variety of stock and bond funds typical of 401K accounts.
- Pension plan fully paid by the company with a *vesting* or wait period of 15 years or when an employee reaches age 65. Employees can retire as early 50 with reduced payout in comparison to age 65. Payment can be arranged in terms of an Annuity with monthly payments or as a lump sum based on the actuarial value of the future payments and estimated life expectancies[41]. For example, employee A has 30 years of service, with a final average pay of $7,000 per month, is age 65 and will be receiving social security,

---

[36] http://www.aon.com/attachments/human-capital-consulting/2017_OG_Major_Integrated_Report_Final.pdf
[37] Ibid
[38] High potential employees are employees who have the ability, capacity and desire to perform at several levels above their current role and are management potential. This concept is covered in more detail on Chapter III.
[39] http://digitalcommons.ilr.cornell.edu/chrr/40/
[40] http://exxonmobilfamily.com/en/finance/savings/your-contributions-and-the-company-match
[41] http://exxonmobilfamily.com/en/finance/pension/payment-options

that person's average pension to be received from the company will be $2,698 per month for life[42].

- Medical insurance benefits, for both current employees and retired employees with a variety of copays, deductibles and maximum out of pocket costs that are among the best in the industry.
- Vacation benefits that increase with years of service with the minimum being 2 weeks reaching 6 weeks with more than 30 years of service[43] and having the option of carrying over to the following year 2 weeks.
- Extensive parental leave policies, with having 8 weeks of *paid time off* for both mothers and fathers after the birth or adoption of a child[44].
- Life and disability insurance, with basic company-provided life insurance of two times yearly salary with options to increase coverage at a cost of up to 5 times annual salary.
- Access to both paid and unpaid *leaves of absence* covering sick time, personal emergencies, as well as flexible work arrangements (telecommuting for example)[45].
- Employee discounts, with ExxonMobil employees receiving discounts of about 10% on certain petroleum products, such as gasoline, diesel, lubricants and others[46]. In addition to this, employees get special employee pricing from other major companies such as Apple products, Microsoft software[47], attractions, wireless phone services (ranging from 10-20%) and other services.

Performance Management

The performance management process in integrated companies tends to be highly structured, with companies defining high level company-wide and business-unit wide goals and then cascading those goals downwards in the organization. There are also goals within each of the function as well that are then cascaded down. Categories or rankings can be numerical such as "1 rating" or wording like "Excellent/Exceeds Expectations".

---

[42] http://exxonmobilfamily.com/en/finance/pension/pension-plan-basics
[43] http://cdncareers.exxonmobil.com/-/media/files/family/resources/benefithighlights2016.pdf?la=en
[44] http://careers.exxonmobil.com/en/new-parental-program
[45] http://cdncareers.exxonmobil.com/
/media/files/family/resources/benefithighlightsflier2016.pdf?la=en
[46] http://cdncareers.exxonmobil.com/-/media/files/family/resources/benefithighlights2016.pdf?la=en
[47] For example, employees can get the latest version Microsoft Office for the discounted price of $20 versus $300-$800 in a retail store. https://www.microsofthup.com/hupus/hup.aspx?culture=en-US

Most if not all companies have adopted a *quasi-forced* ranking system that stipulates that only a certain percentage of the workforce can be a number 1, or number 2 and so on. The ranking process is also adjusted for the pay grade that each employee is at so that expectations for a paygrade 1 would be different than a paygrade 3 or other. Another impact of the performance management process is that the ranking sessions are *rolled up* all the way to the top of the organization where final ranking or *calibration* sessions are held. One of the consequences of these calibration meetings is that certain goals or positions that have more visibility across different groups will have a higher assessment than more localized goals or positions.

Independent Downstream or Pure Play Refining Companies

Independent Downstream companies, also known as pure play Refining companies, derive most of their revenues, earnings and cash flow *primarily* from refining crude oil into valuable products such as motor gasoline, diesel, jet fuel, and other products. These companies may also have wholesale and retail fuel marketing assets. Historically, the refining & marketing sector of the industry has been quite volatile, and a few of these companies have pursued growing their other businesses, such as chemicals, midstream or lubricants. As an example, two companies, Marathon Petroleum and Phillips 66, joined this category of companies, which were spun-off from their integrated companies in 2011 and 2012 respectively.

Independent downstream companies, as the name implies, have operations in the downstream sector of the oil & gas industry. Downstream companies can be generally described to have the following characteristics:

- Refining assets represent their largest operations in terms of capital employed, revenues and earnings.
- Earnings are subject to the cyclicality of refining margins and can vary substantially[48] from *quarter* to *quarter* and *year* to *year*.
- Generate substantial amounts of total cash from operations for the company and may invest this cash into other *higher-margin* areas, such as midstream, marketing, or chemicals[49].

---

[48]www.eia.doe.gov/pub/oil_gas/petroleum/analysis_publications/petroleum_issues_trends_1996/CHAPTER7.PDF
[49] http://investor.phillips66.com/investors/news/news-release-details/2014/Phillips-66-Announces-2015-Capital-Program/default.aspx

- Many refining companies have leveraged their transportation (pipelines, terminals, barges and other forms of transportation) assets into forming Master Limited Partnerships or MLPs to capture higher market valuations[50].
- Devote a *significant percent* of cash flow from operations to shareholder distributions (both dividends and share repurchases).

Career Progression

The career progression philosophy depends heavily on multiple conditions. First, the ability to progress depends heavily on the vintage or type of company. Did the Downstream company come about from being spun off from an integrated oil & gas company, like Marathon Petroleum or Phillips 66? If so, the career processes progression and *how* these companies view talent management correspond more closely to an integrated oil company.

Career progression in general for these companies tends to be highly prescriptive, particularly with those positions impacted by labor regulations, such as operation & maintenance workers in refineries or even engineering. For example, to become a refinery plant manager takes several years of experience in an engineering and operations field, with prior refining experience being in many cases mandatory because of the complexity and safety impacts of refining operations. In a downstream company it would be very rare or near next to impossible to have a non-engineer refinery plant manager, while in an E&P or midstream there will be a higher possibility of having a non-engineer be the top level manager (in fact, in midstream many asset managers are often personnel with a business or trading background, not with an engineering background). Under the risk tolerance section, we discuss in more detail why there's a low tolerance for risk in these companies. Lastly, because of the strong emphasis on *years of experience* versus *pure skills* and *ability* for certain positions, it is many times difficult for various *inexperienced* high potential employees to climb up the corporate ladder faster than expected. This consequence has both *benefits* and *downsides* depending on which point view is taken.

An additional angle is that depending on the function the years of experiences might be some times overlooked for pure performance. Areas that may be more flexible include:

---

[50] http://www.marathonpetroleum.com/News/News_Releases/Press_Release/?id=1655082

- Commercial, Business Development, Marketing & Trading operations, where performance is measured in terms of profit & loss versus simply years of experience. A young, high performer employee has higher probability of having a faster career progression in one of the *profit center* functions than other areas.
- Certain areas of Information Technology and Research & Technology, where sheer potential, ability and fast learning might carry more weight than traditional *time-in-grade* requirements.
- Communications, HR, and other non-technical functions might be more open to having newer *less experienced* employees but with higher potential.

Career Stability

Similar to integrated oil companies, downstream companies offer significant career stability compared to other companies, like independent E&P companies. Career stability in downstream companies depends heavily on several factors:

- Location of downstream assets, with companies that have assets in *economically* or *logistically* challenged areas, such as refineries in the Caribbean, Southern Europe or California, where margins are lower, having a more *unstable* career than other companies. Having a career in a company with challenged assets will be significantly more difficult than with companies in improved areas, such as the U.S. Gulf Coast, Singapore or Western Europe. Job seekers should review each company's annual report to better understand where the company operates and where future career opportunities for today and the future might be located.
- Depending on the career function, certain careers will have more or less stability than others. Generally due to the complexity of downstream accounting, level of automation and other factors many companies in this field have chosen to keep support functions based in their home country of operations. In fact many of the smaller refiners keep their entire support functions on-site with operations.
- Company culture and the level of centralization of decision making. Does the company behave like a *truly merchant* refinery, being efficient and allowing decision making to take at the right level within the organization? Or does it centralize every decision, no matter how small, to the corporate headquarters?

- Composition of assets does the company have operations in multiple businesses within the downstream space or does it really heavily on one sector (i.e. Refining)?

Culture

As mentioned on the career stability section previously, depending on the history of the company, a downstream company will tend to have a different culture. Here are a few questions to ask to better understand the company's culture:

- Was the company formed or spun off from another's company assets, for example like Marathon Petroleum or Phillips 66? If so, the company culture is more similar to its prior owners, since changing a company's culture can take several years or even decades.
- Was the company formed through a recent merger, like for example Andeavor, which was a merger between Tesoro and Wester Refining? If so, the company has a blended culture of the two companies and even of the previously acquired assets.
- Has the company grown fast through acquisitions and acquiring different assets from several regions? If so the company has a blended culture arising from the different heritage companies.
- How does the company approach career or talent management for different functions?
- In terms of career management are *profit center* functions, like Commercial, BD, Engineering, treated the same as *cost center* functions, such as IT, HR or Accounting?

Risk Tolerance

Downstream companies tend to have a high aversion to risk, which stems from several factors:

- Downstream largely being a *margin* basis, whereby cost control has a strong emphasis. Refining earnings can fluctuate widely day to day and quarter to quarter, making cost control a high priority in order to navigate through the volatility in earnings.
- Highly regulated business with a very strong emphasis on operating excellence, requiring that almost every step in the value chain be evaluated and with risks mitigated. In addition to this, the downstream business is impacted by many environmental and

safety regulations, from the EPA, OSHA, Department of Transportation, Department of Labor and many other agencies, which require having a high grade of risk assessment and mitigation in place. For example, many positions in refineries, terminals, and other assets have strict regulatory requirements as to the qualifications and certifications. Usually for any person to work in a refinery in the United States, they must have a security clearance administered by the Department of Homeland department called the Transportation Work Identification Credential[51].

- A significant percent of refineries around the world, particularly in the US, are unionized with extensive impacts and requirements for both contractors and employees that may desire to work in that asset.
- Compensation costs are a significantly higher percentage of total operating costs versus upstream companies, even in a low price environment. The heavy focus on managing costs tends to place pressure on salary for employees versus peers in the upstream world. As a percentage of total costs, compensation costs tend to be among the lowest total costs in upstream, with capital expenditures being number one by far.

All of the requirements mentioned before impact career development in several ways, but primarily in establishing various "time-in-grade" requirements and years of experience that in other industries may not be as common (i.e. Tech industry).

Recruiting

Similar to other large companies, recruiting for most new college hires takes places primarily in the Fall semester with also some hiring and internships during the Spring semester at major US universities. Depending on the company, many universities are selected as "top tier" universities, those whereby the company would have a higher percent of job openings reserved or available for than other lower tier universities. In the oil & gas industry, the majority of university recruiting is done through a combination of function representatives plus HR members. Functions such as Accounting, Engineering, IT and others would have recruiting team whereby they send members to scout for talent at various universities.

---

[51] http://www.marathonrefinerycontractor.com/_Catlettsbur/Step_by_Step_Requirements/Step_3_TWIC_Card_Process/

For experienced hires recruiting is commonly done through several ways:

- Many downstream companies prefer to use *headhunter* or talent scouting firms to hire for new hires. These headhunters or recruiters would bring in several candidates and provide the company to choose from qualified candidates.
- The traditional web application represents a very low percentage of actual hired employees in companies, the same through the use of LinkedIn or other networking sites.
- Hiring away from competitors, such as Valero hiring a highly experienced employee from Citgo or other refineries.
- The use of contractors with the options to hire through staffing agencies for specific functions.

Risk Involved

There are several risks involved with working for a downstream company:

- Highly cyclical business with volatile earnings, with corresponding *hiring booms* followed by hiring freezes and even layoffs.
- High emphasis on cost control, reducing relative compensation in good times to E&P and even Midstream companies.
- The refining business, particularly in OECD countries, is subject to ever increasing regulation and taxation, with fuel taxes accounting for 80% of the end-user cost transportation fuels in most European countries.

Performance Management

The performance management process is similar to other large companies with typical steps as follows:

- Next year's organization goals, objectives and plans are completed from Q3 to Q4 and are then rolled out to the organization's senior managers of all business units and functions. The company wide goals for the following years are usually broad and may include a variety of safety, environmental, financial, cost management and other areas.
- These senior managers and function leads then develop and adapt these goals further to fit their respective groups' specific criteria. For example, for the Chief Information Officer, say that role is giving a challenge of reducing costs by 10%, and then the CIO would have to cascade to his or her direct reports and translate that

number into actionable items. For example, let's say that to achieve this improvement in costs, the IT workforce has to be reduced by 500 employees.
- Middle management then further refines these goals all the way to front supervisors. At the end of performance objective, the individual contributor is either given his or her goal, or has to come up with goals that fit into the areas of focus for the year.

Similar to the Integrated Oil Companies, these companies in the downstream would tend to use performance ranking criteria similar such as "Always exceeds expectations", "Mostly Exceeds" or wording like that to convey the *relative performance* of that individual in respect to their peers. Do keep in mind that performance assessments are mostly done based on several areas:

- Individuals' years of experience versus peers. For example, a *stretch* goal for a new hire employee is definitely not the same for a highly experienced individual.
- Comparison against individuals within the same salary band or paygrade. For example, "Exceeds expectation" means a completely different assignment for an entry level person than for a senior executive.
- Potential, relative to that individual's potential performance, how is he or she performing? Are they stretching their knowledge, bringing in value and becoming a future leader?

Independent Exploration & Production Companies

Independent Exploration & Production companies range from large companies such as ConocoPhillips or Anadarko Petroleum to smaller companies that are publicly traded on the world's stock exchanges. As the name indicates, these companies *primarily* participate in the Upstream or Exploration & Production sector of the oil & gas industry and do not have *significant*[52] downstream operations.

Small E&P

These are companies that produce less than 10,000 barrels per day of oil equivalent (combined natural gas and oil production or BOEPD) and usually have less than 500 employees. Ownership in these companies is

---

[52] Many independents may have some exposure to downstream and midstream operations, for example, OXY, but by no means it is considered their "core" business.

often private and a handful might trade on the stock exchange as small cap or *micro-cap* companies.

Company Culture

A small E&P company culture depends heavily on several factors such as size of the company, where assets are located, heritage companies' culture and management. Companies headquartered outside major oil & gas cities such as Houston and Dallas would tend to have a more small company culture than those in those cities. There are many cultural items that characterize a smaller E&P companies such as higher risk tolerance, high percentage of on the job training, absence of detailed procedures and positions tend to be more exposed to *breadth* and *depth*[53] than other companies. Another area that is characteristic of small E&P companies is the fact that ownership is quite visible and direct since most companies are privately owned and operated. Since many of these companies are not traded in the stock exchanges they have more flexibility in the way they conduct operations, they hire employees and the way financial reporting and controls are structured. For example, a petroleum engineer in a small E&P company would tend to help with all the tasks required to drill, complete a well, build infrastructure, calculate the reservoir and estimate hydrocarbon reserves, help in asset reporting, perform procurement functions and many other more tasks than the traditional role segmentation in the larger companies such as "drilling engineer", "completions engineer", and so forth, which tend to operate more within a silo.

Career Progression & Stability

Career progression tends to be faster than most established companies due to the nature of the business. Another area critical in career development is the fact that roles tend to be more *end-to-end* than focused on a small piece of the business. It is not uncommon to find in these small companies one person that would wear "multiple hats" at the same time, such as bookkeeping, financial reporting, procurement, human resources and many more. One of the key characteristics of somebody wanting to work in these small E&P companies is the ability to learn fast, adapt quickly and be able to do multiple roles at the same time. In contrast to larger companies there is less risk of *hyper specialization* and more of a risk of becoming a true *generalist* with enough detail to understand how the entire business fits together. Another advantage of working for these smaller companies is that

---

[53] Breadth vs. depth concept is covered in more detail in Chapter III, in the *Breadth vs. Depth* section on page 70

most companies are not *credentials-driven* in comparison with the larger companies, so in many positions a high school degree may be all that is needed since most if not all of the learning is conducting in a *sink* or *swim* approach with very little formal training performed.

Career stability, because of the nature of how cyclical the exploration & production business is, positions within a small E&P are the *least stable* in terms of job security than any other companies in the oil & gas industry. There is a distinction worth mentioning between the different types of companies, with companies that have a more stable and diversified asset base would tend to have more stable career prospects. Another criterion that is positive for companies in this category is the fact that many positions are *less prone* to *offshoring* or *outsourcing* than at larger companies. This additional benefit from a career perspective is gained because of the complexity of the oil & gas business, with employees *wearing many hats* in a sense and that co-location, agility and resourcefulness are characteristics that are highly valued over *process work* at larger companies.

Risk Tolerance & Risks Involved

Small E&P companies tend be amongst the highest risk takers in the entire oil & gas industry. Because of the initial risks taken, the rewards also tend to be some of the highest in the industry.

From a career perspective there are a few risks involved:

- Unstable earnings, particularly in the *lean years* in the oil & gas business cycle. One year bonuses might hit an all-time record and next year a company might have employee layoffs.
- Lack of *formal* training versus lots of *on-the-job* training, usually in a *sink or swim* approach, which although overall positive, the lack of formal training may lead to not following the best practices or use the latest and best technologies.
- Long hours and lack of formal benefits when compared to the larger companies.
- Lack of a formal career ladder or expectations when compared to larger companies.

Besides all the business cycle risks inherent in the oil & gas, small E&P companies have additional risks such as:

- *Performance* risks, whereby a few unsuccessful exploration or development wells or even a safety or environmental incident can wipe out or bankrupt the entire company.
- *Geographic* risks, since these companies lack geographical diversification, particularly in a low price environment, if the company's production is concentrated in one geographical location that presents challenging economics (i.e. low netback prices[54]) the company might have to close its door or be acquired by a bigger competitor.
- *Financing* and *interest rate* risks, where a company might simply run out of capital to continue to expand production or if financing is not available at attractive interest rates, then the company might have to divest of assets or lay off employees or simply shutdown.
- *Acquisition* risks, which depending on how compensation is structured may be more of a benefit and an advantage than a risk. From a purely employee's perspective, an acquisition might be a career destabilizer, particularly if a much larger company acquires the company and work is transferred to service centers where the original employees might not be hired on to the larger company.

Recruiting, Compensation & Benefits

Recruiting processes, when compared to larger companies, are usually much more informal and generally are not advertised through the internet. Many of these companies rely on an extensive referral networks or "word of mouth" system to hire employees and may not have an official recruiting program. The interview processes at these companies is not very structured and might simply entail calling their office and seeing if they have an opening in their company. If these companies advertise a position opening, it is typically advertised through association's websites, such as the Independent Petroleum Association, regional chambers of commerce or even at a local newspaper with few companies having *Applicant Tracking Systems*[55] like most medium to larger companies. One of the challenges in applying for these companies is *actually finding them* since they are so widespread all throughout the country. For example, in a recent Texas

---

[54] Netback price is the calculated *net proceeds* received by a company from the sale of hydrocarbons *netted back* to the well. This netback price factors in all disposition, marketing and transportation costs associated with selling hydrocarbons to final consumers and can thought of as a *net revenue* per well.
[55] Application Tracking Systems or ATS are discussed in more detail in Chapter IV in page 103

Railroad Commission Operator report, 70% of all oil & gas produced in the state of Texas is produced by roughly 32 companies[56], with the remaining 30% of production being produced by more than 11,500 companies[57].

Compensation tends to be more volatile and less consistent than more established larger companies, with salary or hourly wage being a small component for all employees' compensation and bonus or shares in the company representing a large payout. In good years, compensation for employees, obviously depending on individual arrangements, would be the *best of the best* with large bonuses and the next year the company may have to cut hours or reduce benefits.

Performance Management
The performance management in and out itself is a lot *less formal* than the companies discussed before in this chapter. Because of less established career ladders or position rotations per se, the performance management tends more of a *binary* decision, whether an employee is a good or bad performer instead of the *relative* ranking found in most Fortune 500 companies. Another item to note is than in most of these companies there's usually no formal *Human Resources* department, but these functions are actually performed by the owner of the company or by just one employee that handles all administrative functions, including accounting, HR, IT or other, so these processes are much more personable and informal than most companies.

Medium Size E&P

These companies produce more than 10,000 barrels per day of oil equivalent (combined natural gas and oil production or BOEPD) but less than 100,000 BOEPD. Examples include Crownquest, Carrizo, and other companies. Many of these companies have transitioned from a small company mindset and have significantly more policies and procedures than smaller companies. Many of these companies have already gone public through an Initial Public Offering or IPO, so they have many additional reporting and compliance requirements than smaller E&P companies.

---

[56] Top 32 Texas Oil & Gas Producers, by the Texas RRC
http://www.rrc.state.tx.us/media/43948/top32producers2017.pdf
[57] Oil & Gas Directory, Operator Contact Information, which lists out all O&G operators that *actively* produce hydrocarbons in the state of Texas http://www.rrc.state.tx.us/oil-gas/research-and-statistics/operator-information/oil-gas-directory-operator-contact-information/

Company Culture

Medium size E&P companies can be considered to still retain some of the startup oil company culture but have made any changes in several areas. Key cultural aspects of medium size E&P companies:

- A significant percentage of these companies are traded on the stock exchanges, meaning that an Initial Public Offering was conducted so that the original investors have monetized their investments to some extent.
- Since they are traded on a stock exchange, they have to perform periodic filings with the securities exchange regulator, such as the Securities & Exchange Commission or SEC in the United States, increasing the financial reporting burdens and internal controls used by these companies. Being a publicly traded company carries a lot more concerns around reporting, segregation of duties and regulatory impacts. For example, no longer a job that has financial implications be performed by the same person, but additional personnel are needed.
- Work will start to get more *fractionated* or compartmentalized into areas and specialization career paths start to emerge. Functions such as "drilling engineer", "completions engineer", or "payables accountant" or "corporate reporting" start to become required as the organization grows in complexity of processes and scope of operations.
- Use of commodity *hedging*[58] practices which protect the company from short-term fluctuations in oil, gas and NGL prices, reduce the upside for the company to profit when prices are going up. This is a particular impact with highly leveraged and indebted companies, with hedging being a requirement by many creditors.
- Geographical diversification starts to emerge, with companies having *divisions* that supervise production operations in one or many more states creating a reporting structure back to the headquarters.
- Organization chart becomes more vertical in nature with a more distant relationship between *owners*, in this case *shareholders*, to *management*, and *employees*. While in a small E&P there was usually one *stakeholder* to be accountable to, often the founder, now a

---

[58] *Hedging* is the practice of receiving a guaranteed price for a particular commodity a company might be selling. Hedging is very common in medium size E&P company which is used as a risk management practice so that the company can focus on increasing hydrocarbon production.

medium size E&P has many more *stakeholders* that you have to navigate through.

Career Progression & Stability

There are many different aspects of continuing a career in a medium E&P company. Career progression in mid-size companies start to become more formal, but it still retains the *can-do, flexible and adaptable* attitude and focus on getting things done with the benefits of more specialized career ladders.

Depending on the company's growth prospects, career progression, when compared with smaller companies actually tends to be faster than their smaller peers. This is particularly dependent on growth prospects, but if a company made it this far without going out of business, the next leg in growth mostly has to rely on internal growth from employees.

Because of the growing size and publicly traded status many position assignments begin to start being divided up due to segregation of duties constraints. For example, no longer you can start finding a "jack of all trades" type of employee, particularly those in key positions that can impact a company's financial statements.

Medium size companies began to have more stability than smaller companies primarily due to the following:

- Geographical diversity.
- Complexity in systems, processes and procedures *increase* employment stability when compared to larger or integrated companies.
- Rapid growth and capital funding from investors accelerate how fast the company can develop acreage and compensate employees.
- Strong emphasis on growing volumes, production and cash flow to reinvest back into the business. Even with accounting losses, investors have enough faith in these companies to continue to invest and grow production at *much faster* pace than their large E&P peers.

Risk Tolerance & Risks Involved

Because companies began to be publicly traded, companies start to reduce their exposure to risks in comparison to smaller companies. No longer is finding the *next best field* at all costs the ultimate goal, but the best exploration finds in a *risk adjusted basis*. More and more medium-size companies at this stage of their evolution begin to create *joint ventures* to diversify away from performance risks and balance their portfolio. Although not as risk-adverse as their large E&P cousins or the large integrated oil companies, these companies no longer have a culture of *wildcatting*, which brings many benefits as well from a career stability perspective.

There are several risks involved in working for a medium size E&P company:

- Business cycle dependencies, although significantly less impacted than smaller companies, medium size companies still face the challenges of being overly dependent on oil, gas and NGL prices. Many well-intentioned companies have endured large layoffs or staff reductions in the price collapse of 2014-2016 with many of these companies simply going out of business[59].
- Despite geographic diversification being more common, a significant portion of these companies are still dependent on a particular area or group of areas that they focus on. If that area or group of areas becomes out of favor from a growth perspective or the company's operations are heavily weighted towards an *out of favor* commodity, for example, natural gas in today's world, then the growth prospects or even viability for the company is still in jeopardy.

Recruiting, Compensation & Benefits

Recruiting starts to become a more formal process when compared to smaller companies, but it is nowhere as structured as the larger companies. Medium size companies often recruit at a variety of venues throughout the year such as job fairs, selective college recruiting, trade schools and many other areas. Many companies still retain the informal process of receiving unsolicited resumes and application with the actual *hiring manager* still conducting the job of sorting through all of these.

Medium companies typically start providing more long-term focus benefits such as 401k matching, improved health insurance benefits but the focus is still on short term compensation such as cash with very few or basically no company offering pension benefits.

---

[59] http://www.tulsaworld.com/business/energy/two-tulsa-energy-companies-confirm-rounds-of-late-february-layoffs/article_a2a4f6c3-a0aa-5b36-8d03-5da5dc2b0ae1.html

An additional area is the fact that most if not all initial investors have monetized their initial investment in the company with many companies not offering any kind of equity-based compensation to newer employees, limiting new employees' upside potential.

Companies in this range start offering benefits such as comprehensive health insurance, 401k matching, but few or none offer pension plans. In general, benefits packages in medium companies tend to be amongst the lowest in the industry. Some companies start offering some form or way of formalized training when compared to smaller companies.

Performance Management

The performance management process at these companies start to get more formal in nature, but in no way still resembles the highly structured, software-driven processes in larger E&P or integrated companies.

Several key characteristics of performance management in medium companies:

- Informal process, which may or may not be documented. If documented, the process might be as easiest as listing out key achievements and contributions in a *Microsoft Word* document or some other basic software.
- The evaluation process is generally not based in terms of *relative ranking*, but instead is simply as "completes the job" or does not complete the job. Since hiring and firing decisions are more fluid than in larger companies, the documentation process is not as heavily enforced or needed.
- The HR department in these companies starts to grow in influence when compared to smaller companies, but it is not as *process driven* as larger companies.

Large Size E&P

In this category we find companies like ConocoPhillips, Marathon Oil, Occidental Petroleum and many others. Large E&P companies can be considered those that produce more than 100,000 barrels of oil equivalent per day. These companies commonly have more than 1,000 employees in total and can even as big as 10,000 employees and have worldwide operations, although the U.S. represents the highest proportion of earnings, production and employee base. Most if not all of these companies are traded on a stock exchange.

Company Culture

As companies start to grow, particularly in the E&P sector, they start to become more and more risk adverse. The culture in these types of companies also depends on *how* these companies were formed in the first place. A couple of questions to think about these companies:

- Were these companies created or formed after they divested their downstream assets, such as in the case of ConocoPhillips[60], Marathon Oil[61] or Hess[62]? If so, these companies still largely retain their original integrated company culture which tends to be much more process driven.
- Did these companies become large independent through sheer *organic* growth or did they grow through acquisitions? If they did through organic growth they tend to hold on more on to their original company culture than those that grew through acquisitions.
- Are the original founders still heavily involved in the *day-to-day* operations? If so, the company would still likely retain the original drive that made it arrive to such as large place.
- Have they adopted major enterprise management software from vendors such as SAP or Oracle? If so, then that company would tend to be more *structured* than peers based on large policy decisions.
- Have the company's headquarters recently moved to a large city such as Houston or Dallas and are they far away removed from the main operations such as Midland or other producing basins?
- Does the company still retain a focus on hiring from within or does it hire from other companies?
- *Cost center* functions such as Finance, IT, HR, Legal and others tend to start to have more *cloud* in the organization than *profit center* functions such as engineering, business development. One of the reasons for what this is the fact that the company is publicly traded and could be a target for litigation when compared to lesser known companies. This is an additional outcome of the shift

---

[60] http://www.conocophillips.com/news-media/story/conocophillips-board-of-directors-approves-spin-off-of-phillips-66/
[61] http://ir.marathonoil.com/news-releases/news-release-details/marathon-oil-corporations-board-directors-approves-spin-marathon
[62] http://www.nj.com/business/index.ssf/2014/05/hess_sells_its_retail_business_gas_stations_to_marathon_for_26_billion.html

towards a risk-adverse organization that frequently looks more *how to avoid* losses than to achieve the biggest gain possibly.

Career Progression & Stability

Career progression becomes more *structured* and *time-based* than medium sized companies; with more of a *centrally managed* career progression than the typical smaller company "sink or swim approach". Career progression also tends to depend more heavily on *credentials* such as an MBA, Masters in Engineering or other discipline than the heavily focused *results driven* mindset of a medium or smaller company.

Careers in the larger E&P began to become more stable when compared to their medium and smaller peers, but are not necessarily immune to layoffs and other job reductions[63]. In fact, depending on the overall size of the company and how fast the company grew during the boom times of the business cycle, many companies in this space are actually *less stable* than some of the medium size companies. This is particularly important for those employees who are seen as *cost center* functions, like IT, Finance and HR, with the perennial *"out of sight, out of mind"* implications.

Risk Tolerance & Risks Involved

Risk avoidance becomes a big part of the company's central operations. Although not as heavily focused on risk avoidance as their integrated oil company counterparts, reducing high impact events becomes a way of life for these companies. In fact, the majority of the portfolio of assets of these companies often arises out of the *Joint Venture* agreements with other similarly sized counterparts to diversify the previously mentioned risks inherent in the oil & gas business.

Along the lines of similarly large companies some of the biggest risks involved in working for large E&P company:

- *Visibility risks*, the ever *out of sight, out of mind* becomes more common place as the direct line between decision makers and employees becomes longer.
- *Process driven* mentality, which reduces spontaneity and quick decision making that characterized the medium and smaller companies.
- *Reactionary* to Wall Street or investor trends, since companies have to report earnings on a quarterly basis and equity research analysts

---

[63] Insert quote regarding COP layoffs

can downgrade a company that is not following the latest investment fad.

Recruiting, Compensation & Benefits

Recruiting in large companies becomes a lot more formalized and process driven than the smaller counterparts. One of the reasons for this is the fact that the company is now larger and it is subject to increased scrutiny in hiring and termination decisions when compared to smaller companies. Recruiting in these companies is typically driven from HR processes and it is highly structured in terms of using recruiting software to sort through applications. Despite the highly structured recruiting process, similar to any company, the best option to apply and get a job is through referrals, either formal or informal.

Compensation for larger E&P companies tends to be amongst the best in the industry offering highly competitive salary, bonus structure as well as world class benefits. Benefits such as *generous*[64] 401k matching, top quality health insurance and other types of insurance. Plus, many companies still provide pension plans of different forms. In addition, stock-based compensation is common for high pay grade employees.

One of the biggest benefits of working for a company of this size is the *high emphasis* on structured training versus other companies. These highly structured training programs offer a good transition, particularly for college hires, than starting with very little oil business hands-on experience and allow these new hires to ease in better to the requirements of a new position instead of the traditionally "sink or swim" approach at the smaller companies.

Midstream Companies – Master Limited Partnerships (MLP)

Midstream companies are primarily engaged in the gathering, processing, transportation of natural gas and crude oil, fractionation of natural gas liquids (NGLs); crude oil, natural gas and refined products pipelines, and crude oil and products terminals as well as other assets. Many midstream companies, especially in the U.S., are organized as a Master Limited Partnerships or MLPs instead of traditional C-corporations. An MLP is a tax-advantaged or pass-through entity for U.S. Federal Income Tax purposes since the entity's income flows directly to each unit holder.

---

[64] Insert quote regarding COP 401k in Bloomberg

Company Culture

The culture of many midstream companies is highly influenced by its commercial and business development nature. For an MLP to grow its distributions and grow distributable cash flows to its unitholders the company needs to consistently acquire new assets, build new assets and sign commercial agreements. Another factor that encourages midstream companies to have a very *commercial-driven* culture is the fact that competition is the highest in the oil & gas industry. In fact, most of the leading midstream companies tend to be heavy in the Commercial BD area and a significant majority of leading executives do not come from an engineering or operations background but more from a Commercial or Trading background.

Key characteristics of most midstream companies' culture:

- Rapidly changing and ever evolving business that emphasizes the ability to learn quickly, apply knowledge and succeed in the market place.
- High focus on commercial agreements, growing distributions and adding or building new assets into the portfolio.
- High job security, particularly for support functions, because of the sheer complexity and lack of industry wide systems. For example, the majority of software vendors in the oil & gas industry target upstream and to a lesser extent downstream companies, but very few have dedicated midstream specific software. This creates a need for highly customized and complex solutions, increasing the learning curve for newcomers and creating a competitive advantage for those already in the business.
- Due to the fact that most companies are MLPs, focusing on distributable cash flow and building long term projects that create steady cash flows is paramount. This is contract with downstream for example where the emphasis is on controlling costs. Because of this focus and the industry need for infrastructure to move hydrocarbons from one location to the other, there's a high need for new talent being added to these organizations.

Career Progression & Stability

Career progression highly depends on the function selected, with some other fastest progression in the *profit center* functions, such as trading, business development, contract negotiation and even some key engineering functions. *Cost center* functions such as finance, IT, HR and many others would tend to have a slower career progression than the profit center functions. In general, one of the biggest selling points of working for a midstream company is the fact that younger or less experienced employees can rapidly gain new skills, learn the business and improve their understanding at a much faster pace than other businesses.

Progression in the commercial space tends to be significantly higher pace than at other functions, with many high potential employees starting as "Marketing Representatives" or "Asset Analysts" then moving to running economics for contract deals and later on leading negotiations with large producers. Again, the commodity that is most highly favored, for example for the past couple of years it has been crude oil, would tend to experience a higher volume of contract deals and new and improved markets while those commodities out of favor, i.e. dry natural gas, would be more of a managing existing agreements versus adding new capacity.

One factor that is highly critical with midstream companies is the type of underlying business that they operate in. For example, companies with high emphasis on natural gas gathering & processing versus pipelines would have exposure to natural gas and natural gas liquids prices. Geography also impacts the career growth and career stability of midstream assets, with assets located in areas with high production growth will have a more stable career path than those assets located in shrinking or declining basins. A current example of a high growth area is the Permian basin in West Texas, where several midstream assets, for all commodities, have been built since the shale revolution began in the mid to late 2000's. This is in contrast to plateau or ultimate decline areas such as the San Juan basin, or even the Gulf of Mexico in the United States.

Risk Tolerance & Risks Involved

One of the unique risks with Midstream companies is that they sometimes can focus on only oil or gas, but not both, particularly in the pipeline business. This can carry significant implications for hiring and employee retention.

However, one of the plus sides is that smaller midstream companies can be done as flips where they build out a system in 2-5 years and then flip it to someone else, like an EPCO, KM, Magellan, etc....If you can work for one of the flipping type companies you can potentially work your way up, because the folks at the top make so much money they do not all return every time. Also, some will make a large payday and then go work for a company with a nice benefits package that is low stress.

Recruiting, Compensation & Benefits

Recruiting for midstream companies depends on the size, with larger companies often recruiting out of major state or private universities for their college hires. Most companies also include hiring from other industries that have similar commercial focus, such as banking.

Because the majority of midstream companies have been formed in the recent past couple of decades (compared to integrated oil companies with 100+ years of history) very few companies offer pension benefits. The compensation in midstream companies tends to be among the lowest for support or cost center type functions while for the profit center functions tends to be amongst the highest in the industry. For these profit center type careers or functions, the preferred method is to provide cash bonuses with a significant portion of equity compensation in MLP units or other types of tradable equity or options. These compensation schemes are geared toward retention of high potential and highly productive employees.

Other benefits, such as health or life insurance tend not to be as competitive as companies in other sectors and typically provide the minimum benefits to retain employees. One of the biggest benefits of working in a midstream company is that career progression is amongst the highest in the industry, with rapidly changing business, addition of new contracts and agreements that impact all functions.

Performance Management

Because of the high emphasis on commercial aspects, the performance management process at these companies is amongst the most informal in the industry companies. Similar to other sectors, the larger the size of the company the more *structured* the performance management process will be in comparison to the smaller companies. However, a good distinction must be made between the profit center functions versus the cost center functions. Performance management processes in the cost center functions tends to be much more structured than for the other functions. For example, an IT employee may have a target customer service satisfaction or percent uptime, which is a highly structured and quantifiable. On the other hand, a commercial representative may have one or two big projects that they work on during the year, for example, the conclusion of a natural gas gathering & processing agreement and as long those objectives are achieved at good terms for the company then that employee would receive the highest rating and have compensation incentives tied to it. Another factor is that the differentiation between employees in the same profit center function tends to be much clearer than those in support or cost center functions. For example, how do you rank finance employee "A" which completed 100 bank reconciliations versus another that completed the budget in time? There's a lot more discretion and subjective valuation in the cost center side than for say a commercial representative who just signed up a $400 million long-term gathering & processing agreement.

Service Companies

Service companies provide equipment and services to upstream, midstream and downstream companies. Services companies, as their name implies do not themselves *own* upstream, downstream or midstream assets, but instead *provide* equipment and services to companies operating in the sectors mentioned before. In periods of elevated industry capital expenditures (CAPEX), these companies *tend* to perform very well. Conversely, in periods of reduced capital expenditures, services companies would *tend* to *underperform* the other companies.

It is impossible to list of the types of services companies and the associated factors to consider when working for them. Here are a few things to remember. All the comments we have made about size and how it relates to jobs with Oil and Gas companies applies here. Also, you might find yourself being more valued by working for a service company than an oil and gas company because you, with the right skill set, could impact sales or income directly for a company. In larger Oil and Gas firms, rarely does one person do this. It is usually teams of people or accomplished through buying out other firms. The downside to working on the service side is that you are never the client. This means that your firm has to excel at taking care of its clients, sales, and marketing.

Company Culture

Service companies tend to have a culture that is more similar to a *manufacturing* and *selling* business than any other business associated with the oil & gas industry. Upstream have very little to zero sales or marketing emphasis, downstream has a significant focus on marketing and sales, but the sales are typically among other oil companies or large customers, while with midstream the focus is on contracts with producers, shippers or other large customers. In the service industry, customers can be a large as RoyalDutchShell to a small as a 1000BPD producer in West Texas. In addition to this, service companies many times deal with manufacturing processes or in some cases may only provide purely services and not sell any equipment per se.

Career Progression & Career Stability

Careers in the service industry tend to be more amongst the highest impacted by the level of commodity prices. When prices recover, service companies struggle to find the right amount of candidates while when prices start to decrease, upstream, midstream and downstream companies cut capex which highly impacts the demand for goods and services from these companies.

Even within the service industry there are different subsectors or areas that are more impacted by the price of oil than others. For example, manufacturers which provide long life equipment such as pumps, compressors, motors and other will do fairly well with a high level of capital expenditures, but those same companies will experience staff reductions when capital expenditures are constrained. Those companies that provide more "keeping the lights on" goods and services, for example, compressor maintenance and repairs, critical low value spare parts manufacturers and production maintenance services will tend to be more stable but have lower compensation.

Risk Tolerance & Risks Involved

Service companies in general tend to be amongst the companies with the most *risk taker* mentality in the oil & gas business. Particularly important is the fact that they have to sell *products* and *services* to companies, unlike the traditional oil companies which have a ready market for their primary commodity business. For example, when a service company decides to launch a new product, it takes several risks that the product may not sell well, R&D might be wasted, and inventory, wages and other components might not be recovered. Since service companies are more of a hybrid between a manufacturing and marketing business the company, by necessity, usually tend to be more risk takers than others.

There are several risks involved in working for a service company:

- Highly cyclical business, even more so than traditional upstream, midstream or downstream oil companies. A product line could become obsolete by the market quite fast and could result in lack

of job security if the company cannot retrain its employees fast enough or shift them to other divisions.
- Dependent on gaining contracts with oil companies that, depending on the particular product line, could have severe competition. Please note that this is highly dependent on the particular product being sold, if a product does not have an equivalent in the market and is one that companies highly need, the employees supporting that product line would have a lot more job security than other. Companies that focus on *commodity* products that are highly interchangeable would have a tougher time with pricing power over its customers versus those that have an *irreplaceable* product that everybody needs.
- Since positions also tend to be cyclical, there is a less other non-cash benefits, such as pensions in comparison to other subsectors in the industry.

Recruiting, Compensation & Benefits

Recruiting for service companies tends to be aligned in terms of functions with more engineering and science-based functions being recruited out of major universities. Other functions and operations will likely be recruited out of smaller universities, community colleges and trade schools. Similar to the research functions in the upstream or downstream business, there is a significant *talent shortage* for highly specialized fields, particularly those that require PhDs and other advanced degrees in geosciences, petroleum engineering, chemical engineering, mechanical engineering that require years and years of recruiting, so the job security for those type of jobs will be higher than other ones.

Similar to midstream, the recruiting process tends to be most structured with larger companies, such as Schlumberger, Baker Hughes, Halliburton and the like while the smallest companies having less of a structured process.

Because of the high volatility in the service industry, the compensation philosophy tends to be geared towards cash compensation with a minimum or standard amount of benefits. Most firms generally do not provide pension benefits to employees and there is a high focus on hiring contractors to provide for short-term staffing needs versus long-term employees. This also varies depending on the function and level of skill required, with PhD and engineering functions being provided more lifelong employment opportunities than more cost center based functions.

Performance Management

The performance management process depends on how large or small the company is. Larger companies would often have more structured, ERP and HR driven performance management processes than smaller companies. Because of the high use of contractors and other contingent workforce resources, there tends to be less of an emphasis on performance management and ranking process. In addition, cost center functions performance agreement rankings and process is typically much more detailed than those in the profit center functions. In the profit center functions for the service companies, such as business development, marketing, and sales, key objectives and goals are usually much more measurable and less subjective in nature. For example, a sales rep who hit a $20 million sales or margin target for the year will has clear *tangible, measurable* and highly *impactful* results for the entire company while an HR or Finance metric is more around *negative* sanctions and avoiding mistakes, errors that could put the company at risk. Commensurably, the career development for those in the profit center functions will likely be targeted, effective and have higher budget associated with training, networking and skills development than those in the cost center functions.

Chapter III – Talent Management & Career Development

"I've missed more than 9,000 shots in my career. I've lost almost 300 games. 26 times, I've been trusted to take the game winning shot and missed. I've failed over and over and over again in my life. And that is why I succeed" – Michael Jordan

Historical Career Developmental

Historically, oil & gas companies have always had challenges looking for qualified employees. Ever since the first commercial oil well was drilled in Pennsylvania, the search for *talented, qualified,* and *willing* employees has been an ever quest. The hiring process has been a key part of the oil & gas industry since the very beginning, with one of the first American oil wells. Entrepreneur George Bissell *hired* famed Yale Professor Benjamin Silliman to prepare a geological assessment of Titusville formation in Pennsylvania[65]. Other notable examples include John D. Rockefeller hiring two of the first chemists, Herman Frasch and William Burton, to figure out how to better refine raw crude oil into different usable and higher valued products. Eventually Rockefeller's team of chemists developed more than 300 by products from each barrel of crude oil[66].

Shell Oil and HAIR system

Many years ago, Shell implemented at the time what they called the HAIR performance appraisal system, one of the industry's most widely used performance appraisal systems. HAIR stands for:

- **H**: High level vision from a Helicopter
- **A**: Power of Analysis
- **I**: Imagination
- **R**: Sense of Reality

This performance appraisal was based on the person's *estimated potential* not on the *actual performance*. This is a key distinction between traditional systems and Shell's system, the ability to rank and evaluate based on the *potential performance* of an individual. The performance process at Shell was conducted by committees around the world that were typically higher than the individual contributor's direct supervisor or manager. Similar systems to Shell's Hair have been implemented throughout the oil & gas industry.

Impact in Government

Shell's performance and talent management were so popular that Singapore adopted Shell's system for scouting at high potentials and implemented it into its civil service system in the early 1980's. This system replaced the British system of managing its Civil Service it had inherited[67]. Shell's focus

---

[65] https://aoghs.org/petroleum-pioneers/american-oil-history/
[66] https://fee.org/articles/john-d-rockefeller-and-the-oil-industry/
[67] Public Administration Singapore-style, by Jon S.T. Quah, pages 79-80

on identifying the *long-term* potential of its employees was well known among multinational corporations and caught the attention of the Prime Minister of Singapore, the late Mr. Lee Kuan Yew and decided to select this system. From an interview conducted in 2009:

> *"I was intrigued at the ability of Shell to have a talent pool spread around the world, over 100 countries, yet pick the right people for promotions"* – Lee Kuan Yew[68]

For him, this was a system worth copying into the Singaporean civil service system.

Traditional Oil Company model

Before the 1990's, large oil companies followed a traditional American company culture in the fact that employees were more or less expected to work for the same employer for their entire career. Although not the same as it used to be before the late 2000's, large, established, integrated oil & companies like ExxonMobil, Shell, Chevron, and Total still follow this model of long or even lifetime employment. It is not uncommon to find employees with 30 plus years of service with the same company.

This has occurred despite the large layoffs that usually follow a substantial decline in the price of oil and natural gas.

For the following companies, their named executive offers have been with their respective companies for more than 25 years, a record when compared to other industries, such as information technology, consumer durables, and manufacturing and consumer staples:

- John Watson, former Chairman and CEO of Chevron worked for the company since 1980[69] and retired in 2018.
- Ben van Beurden, CEO of Shell, joined the company in 1983[70].
- Ryan Lance, CEO of ConocoPhillips, joined predecessor companies in 1985[71].
- Gary R. Heminger, Chairman and CEO of Marathon Petroleum joined Marathon in 1975[72].

---

[68] https://peoplecentre.wordpress.com/2015/01/16/succession-planning/
[69] https://www.chevron.com/about/leadership/john-watson
[70] https://www.shell.com/about-us/leadership/executive-committee/ben-van-beurden.html
[71] http://www.conocophillips.com/about-us/leadership/ryan-lance/
[72] http://www.marathonpetroleum.com/About_MPC/Corporate_Profile/Corporate_Officers/Gary_R_Heminger/

Impact of Mergers & Divestitures

In the wave of merges starting in the late 1990's and continuing until the early 2000's, mergers have had a significant impact in careers, particularly for the major oil & gas companies. Usually, in a merger transactions, the acquiring companies' senior leadership stays in place while as part of the negotiation, the next layer of management would generally come in from the acquired company. Results of these can be found in some of the major IOCs like Chevron where the top management came from Chevron while the next layer of management was brought in from Texaco.

Specialist vs. Generalist

To specialize or not is indeed the $1 million question! Although this dilemma impacts other industries, it is quite prevalent in oil & gas, particularly due to the complexity and extensive use of third party systems and processes in a company. There will always be a need for highly specialized fields, such as geoscientists or petroleum engineers, the question remains for the individual whether they would like to progress to management or remain a specialist.

Let's define terms first; a generalist is somebody "who knows something about a lot of subjects"[73], or in other words a *"jack of all trades"*. Specialists are typically the experts on that particular field and tend to have very defined roles. For example, a specialist might have a title like "Rotating Equipment Subject Matter Expert" while a generalist might have a more general or generic titles like "Enterprise Asset Management Director".

In fact, the world relies on constant specialization and specialization is one of the key components that has made the world richer. In economics this is called the "division of labor", a term fully described by Adam Smith in his *"Wealth of Nations"* book in the 18th century. In his book, Adam describes the ability of a pin factory to increase output if the *full task* of producing a steel pin was *decomposed* into more manageable *individual tasks* completed by different people. This division of labor today is what allows the economy to grow and prosper by having people focus on a small set of tasks within the economy and thrive. Not many companies in the world are good at every business, which is why there is a need to specialize. The same applies to individuals in oil & gas, knowing about everything is very difficult, in reality an impossible task, therefore needing specialization for that very reason. The book *"I, Pencil: My Family Tree as told to Leonard Read"* was written on

---

[73] https://www.cleverism.com/ultimate-career-choice-generalist-vs-specialist/

this very subject[74]. This book details the advanced processes required to produce something as simple as pencil. For example, for the wood required to manufacture a pencil you need workers that would cut down trees, process the wood into small pieces that can be used for making a pencil. In the entire process individuals or companies who do not even know themselves work in synchronization through the price system to eventually create a pencil. The conclusion of the book is for a modern economy or company to function, specialization is essential. The main idea of this essay is that people are limited to what they can do. The pencil talks about what it is made of to show all the people that are needed to make it. Unfortunately, none of these people actually *know how* to make the pencil all by themselves. The pencil is created because of natural forces that cause one to do what they know in exchange for other goods/services[75].

Career Progression

Generally, earlier in your career, you will probably be asked to *specialize* and learn everything about a particular area, system, process or function, but as you progress throughout your career, being a generalist will be more important. Many companies also provide two career ladders, one for specialists or Subject Matter Experts (SME's) and one for those in management tracks. One is not necessarily better than the other; choosing one over the other is based on personal preference, and long term expectations. One of the key skills that typically sets apart a generalist vs. a specialist is communication skills and the ability to communicate *complex ideas* in *simple terms*, as well as the ability to sell a product or solution.

One of the hallmarks of career progression, particularly for those pursuing a management ladder, is the first supervisory assignment. This is one of the critical assignments in an individual's career since it marks the transition from an *individual contributor* to a supervisory role, where *significantly different* skills are needed in order to succeed. Many employees fail to make this successful transition since a supervisor has to accomplish results *through others* instead of being achieved *individually*. Successful supervisors and managers are often those that can *delegate* and *mentor* employees so that they can achieve results for the larger group or department.

Advantages of Specialization

There are many advantages of being considered a specialist in a company:

---

[74] http://www.econlib.org/library/Essays/rdPncl1.html
[75] http://acnhowtobeyourownboss.blogspot.com/2008/10/i-pencil-summary-quotes-and-analysis.html

- Higher job security in the *short* to *medium* term. Although in the long term, specialization carries risks of absolute disruption, such as a when system or process becomes obsolete, the value of specialization is reduced since that system or process is no longer needed.
- If you are interested in knowing everything about a specific topic, then your personal preference would lean towards becoming a specialist.
- If you enjoy producing independently and do not like to delegate tasks or assignments, a specialist role is best.
- Since you are considered an expert, you have to spend *less time* selling an idea if you are widely seen as the SME on that topic than a non-SME would. In other your input carries more weight because of your experience.
- High reliance on *hard* skills, like knowledge of a particular software, tool, or skillset with normally less development of communication or social skills.
- Depending on the area being specialized, higher short to medium earnings if your area is in high demand, for example highly specialized reservoir engineers in a high oil price environment can easily switch from one company to the other because this specialization is in high demand.
- Higher compensation, again if the demand for that field is growing.
- You are generally tied in more to the success and demand of your *particular specialization* than you have to be reliant on your *particular company*. If you have very valuable skills in high demand and your company goes out of business, it is highly like you will have a job within a reasonable timeframe. On the other hand if you are a generalist and your company goes out of business, you may have a more difficult time finding a job quickly.

Disadvantages of Specialization

- Narrower career path and fewer opportunities to grow. A company can only have so *many* narrowly defined experts in a field.
- Fall into what is famously known as the "curse of competency". This concept is discussed later in this chapter.
- Over time, a specialist would tend acquire less *general* or *broad* knowledge and how everything fits together which in times of rapid change might be needed.

- Less potential to move into senior or executive management, typically all if not *most* of the senior leadership of a company are generalists, and very few are considered an SME in a particular topic.
- More reliant on overall demand for your specific area instead of being dependent on the overall health of your company. In other words, let's say your company no longer uses particular software or has exited a business altogether, your services might not be required by that specific company.
- By definition less adaptive to *disruptive change* than generalists.

Advantages of being a Generalist

There are several advantages of being a generalist, among which are:

- More career opportunities and the ability to have more *horizontal* and *vertical*[76] career assignments.
- A generalist becomes the *point of contact* or *integration point* between the different SME's in a company, with generalists usually being assigned into project management roles.
- Usually a generalist would be more likely to see the big picture and not get bogged down in the *minute details* that may not necessarily impact the decision in the long run.
- High reliance on communication, management, and other soft skills over hard skills.
- More transferable skills, particularly if there are difficulties in your company or the sector you work in becomes out of favor. For example, in case of a collapse in oil prices, a generalist employee in an oil & gas company would have a higher change of finding a position than a highly specialized geoscientist or reservoir engineer.
- Higher changes of being promoted into executive or senior management.
- More adaptive to change, particularly *disruptive* and *unexpected* change.

---

[76] Horizontal refers to a latera or a move that involves a different skill but does not generally entail a promotion per se. A vertical move usually implies a promotion and involves becoming a supervisor

Disadvantages of being a Generalist

- By definition, a generalist sacrifices *depth* over *breadth*. With higher and more complex processes in the oil & gas industry, this is a considerable point of concern.
- Less *short* and *medium* job security than a specialist. In case of a downturn, it is more difficult to assess the value or output a company can get out of a generalist vs. a specialist.
- High reliance on *soft skills*, like presentation, communication and ability to gain report and be charismatic. If you are somebody who prefers not to do presentations, have trouble communicating ideas, then the role of a generalist might not be the best fit for you.

Cost center vs. Profit Center

Another factor influencing career development depends on whether one's function or department is considered a *cost center* or a *profit center*. Cost centers are typically *back office* functions that do not directly bring revenues to a company. Example of cost center functions in an oil & gas company includes finance, IT, HR, engineering support, maintenance and others. Example of profit center functions in a typical oil & gas company are supply, trading, scheduling, corporate and business development, marketing, pricing and other areas.

Profit Center

Keep in mind that the definition of what groups or departments constitute a profit center can vary from sector to sector. For example, for an *Engineering, Procurement & Contracting* or EPC firm, the engineering department will generally be considered a *profit center*. But for an *operating* oil & gas company, the engineering department will be considered a *cost center*.

Functions or departments that have profit & loss impact often have:

- Higher overall budget and resources.
- Higher compensation & bonus structure.
- More opportunities for rotations *in* and *out* of several businesses units.
- Higher opportunities for external conferences and other training, since budgets are overall higher.
- Faster career progression with assignments that impact a company's P&L directly. A younger employee that increases P&L

by 400% can be more rapidly progressed through the ranks than an employee working in a cost center type function.
- Companies are willing to pay higher salaries and bonuses and pay for top performers since they bring in additional sales, or can close additional Business Development deals.
- Less susceptible to *outsourcing* or *offshoring* since these positions are client facing and thrive on interaction with outside customers or clients.
- Easier to differentiate performance from one individual to the other since sales or business development metrics are easily quantifiable, and can be associated to a particular individual or group.
- Better alignment between effort and compensation, with bonus structures fully aligned in bringing additional revenues or deals.
- *Positive sanctions*, with thinking outside the box being more highly welcomed.
- Less risk aversion than cost center type functions.
- More visibility into executive management.

Cost Center

Similarly to the definition of a profit center, a group that would typically be considered a cost center in an operating company, such as IT or accounting, could also be considered a profit center for an Accounting or IT consulting firm, such as the Big Four firms[77], Accenture and many others.

In a cost center type function there is a different emphasis and set of incentives that impact behavior:

- Lower overall budgets, which impact average compensation and bonus structure.
- Emphasis is on *reducing costs* and not *bringing in* additional revenues like in a profit center department.
- Slower career progression and career ladders based on *time in job* and not overall performance.
- Can be conducive to have an *out of mind out of sight* mentality, which makes a cost center be subject to employee reductions, *outsourcing* or *offshoring*. Less overall visibility than a profit center would have.

---

[77] The Big Four accounting firms are Deloitte, PWC, EY & KPMG

- Often experience *negative sanctions* where behavior is usually penalized such as a person making an error versus learning from that mistake and thinking outside the box.
- Excessive risk avoidance due to very few *positive* sanctions being encountered.
- Very difficult to assess value bringing in to the company in comparison to a profit center group. No direct correlation between performance for the company and to the individual or group.

Curse of Competence

The curse of competence is typically defined as "when a person is talented, recognized as a great performer, and rewarded with... *more work*"[78]. In other words, being *too competent* can restrict your career opportunities. Throughout history, workers have found themselves in the position of being so valuable to an organization that they lose out on opportunities to advance[79]. As summarized in a recent article *"The competent gets to do everything"*[80], is a reality for many people in the oil & gas industry.

There are many advantages of being considered the *go-to* person in a group or department, but there are also many *disadvantages* as well that come along with that.

Advantages

Some of the advantages of the "curse of competence":

- Higher job security due to the fact that you are being relied upon and recognized by multiple people in your organization.
- Predictable work which allows for higher work & life balance than a position where you are constantly struggling just to catch up. This would depend on the workload being placed on that highly competent person.
- Internal satisfaction from doing a good job and feeling valuable to the company.
- Working in a field where you are seen as the expert can improve your credibility in a subject versus somebody with less competence.

---

[78] http://thedayjob.bangordailynews.com/2016/02/23/home/the-curse-of-competence/
[79] http://www.businessinsider.com/being-too-good-at-your-job-could-actually-hurt-your-career-2015-7
[80] https://www.mrgazz.com/writing/essays/the-curse-of-the-competent-mainmenu-83

Disadvantages

Some of the disadvantages of the "curse of competence":

- Higher expectations since you have been proven to succeed in the past, in other words, you will get more work assigned to you *just because* you can handle it.
- Higher probability of getting stereotyped in a specific role or skillset. For example, if you are highly proficient with a particular software, everybody will associate you as the "go-to person" for that particular skill.
- High probability of underestimating the time it takes to complete work, since everything looks easy on the outside, particularly for those not doing the work, there will be a higher probability of getting work assigned with *unrealistic* deadlines.
- Less development of other skills that you might need if you are interested in progressing up the career ladder, such as ability to change and ability to take on new challenges.
- Less ability to move up since so many people in your department or area depend on you to keep things running. People that are promoted faster can easily transition into other roles and not have a significant impact on the department or group.
- More resistance to change and a higher expectation if an employee is transitioned into another role. *What happened to Joe once he transitioned into Sales? He used to walk on water in accounting!*
- Inability to grow and develop new skills, since your department or group will rely on you for more and more critical items. You would have less time to develop other critical skills for your long-term career.
- The more indispensable you become, the less probability of being promoted or transferred to another group.
- Less outside of the box thinking as those with the curse of competence have been doing that particular skill or activity for many years.

A promotion is not based on your past performance, but on *your potential for future work*.

Several signs you are indispensable

Although the common saying goes that nobody is considered indispensable, we will list out several signs or indicators that would point to your current position being labeled as *indispensable* (in the short term at least):

- Does your boss call you or relies on you as his or her *right-hand* person?
- When you take vacation, does the regular flow of business have to be halted and wait until you come back?
- When you take time-off does everybody email you asking for help?
- Does everybody know or sense when you take time off?
- Whenever there's a problem, is your name the first one that comes to mind?
- Are you considered a rock star but do not progress through several positions within a reasonable time period?
- When was the last time you got promoted? Did your promotion entail learning a completely *new skill* or does it build on the same skills used before?
- Have you had the same supervisor or manager for several years now?
- Does your boss always mention that you make his or her life so much easier?
- Do your colleagues or boss have unrealistic expectations as far as how fast you can finish an assignment?
- Does your position or assignment no longer seem like a stretch?
- Are you developing new skills, particularly *soft skills* and other management skills in your current position?
- Is the cost of replacing you *too high*?
- Are you stereotyped as far as your skillset? For example, when all your colleagues think about your name, do they think about 1-2 skills or abilities or multiple abilities? If they think about 1 or 2, then you are more likely to be going through the curse of competence.

Practical Advice on how to evolve out of the curse of competence
There are several recommended steps or strategies to evolve out of this current state:

- Discuss with your supervisor or person responsible for career development what your next move or position would look like. This is a particular applicable question if you have been in your role for several years now; it is not advisable to ask this question if you recently have been moved to a new position.
- Network within your company and volunteer for an assignment completely out of your comfort zone. For example, if you work in accounting maybe assist somebody in trading or marketing to see if that would be an activity or function you would enjoy doing for your next assignment.
- Even within the same position, volunteer for assignments that stretch you, such as public speaking at a company event, or volunteer for employee groups.
- Find a mentor outside your immediate department that can provide you with feedback on how to develop new skills in a function outside your experience. Growth requires being uncomfortable, not being able to have all the answers right away and have the ability and willingness to learn.
- Network outside your company to see if this curse of competence is typical for your particular sector or particular function. Certain *cost center* functions, such as accounting, IT and HR commonly tend to have more of this problem than other functions, so it might not necessarily be the company you work but the *function* or discipline that you work for. Consider a change of function within the same company.

Ability to adapt

The ability to adapt or change is one of the key characteristics senior managers look for in high potential employees. Just looking back 20 years in the oil & gas industry here are just a snapshot of events that have changed and impacted this industry:

- In the mid to late 1990's, increased production from new heavy oil projects in Venezuela and Canada and associated refinery upgrades in the U.S. Gulf Coast and other regions to process these heavier oils.

- The advent of automation, *offshoring* and *outsourcing* of back office processes, particularly in accounting and IT. Many companies prior to the 1990's had *in-house* and embedded back office support.
- Oil production from deepwater and *ultra* deepwaters from many regions in the world.
- Massive increases in LNG capacity around the world and the U.S. switching from becoming an *expected* LNG *importer* to becoming one of the largest LNG *exporters*.
- Lifting of the ban on exporting U.S. produced crude oil.
- The shale revolution, which caused the U.S. to swing from the largest importer of both raw crude oil and finished petroleum products to now becoming a swing exporter of light crude oil and overall *net exporter* of refined products.
- The renaissance of the U.S. petrochemical and refining industries due in large part to the shale revolution, which allowed to produce crude oil, natural gas and natural gas liquids in unconventional reservoirs at an unprecedented production rate.
- The building of pipelines, processing plants and other infrastructure in the Permian basin, a petroleum basin that as early as 1999 described as being in a plateau and going through a *structural* and *irreversible* decline, with production having peaked in the mid 1970's. Today the Permian basin is the largest producing region in the United States, producing close to 626MBPD of crude oil[81].

In other words, any industry, but in particularly the oil & gas industry, should be thought of as an *ever changing, ever evolving* industry. Companies that do not adapt to new changes or cannot set up a *differentiated* strategy versus its peers will simply not survive in this new landscape.

As mentioned before, employees with *high capacity* to adapt quickly can benefit in several ways:

- Get onboard quicker with coming changes and understand how these changes will impact the employee's company.
- Be able to communicate the upcoming changes to their direct group and increase understanding and *buy-in* from co-workers.

---

[81] https://www.eia.gov/petroleum/drilling/pdf/permian.pdf Data as of April/May 2018.

- Be seen as a *change agent* in your company that *welcomes, embraces* and *manages* change instead of avoiding it.
- Since many changes are outside of our control and more likely inevitable, welcoming change can be seen as a breath of fresh air instead of actively avoiding it.

Here are ways managers look for the ability to adapt or change in an individual:

- How does the individual do when transitioning to a new rotation or assignment?
- Is the employee comfortable with change and learning a new business, process or way of doing things?
- More important, is the employee OK with going from a *high level* of competency and expertise in a particular area and start from *scratch* in a new area? This is one of the categories where a significant number of employees *fail* to transition effectively, since they are transitioning from an area where they have a high level of proficiency to a new field where they are the least experienced team member.
- How does the employee handle going from being the *smartest person* in the room to being the *newbie*?

With change or a new way of doing things there is always a learning curve and reduced productivity at the beginning. For example, try to use your computer mouse with your *non-dominant* hand and see how that goes? More than likely your productive will take a decline in the short-term. With change there is definitely a short term *trade-off* but typically we see higher benefits in the medium and long term.

Breadth vs. Depth

Breadth vs. depth is a key question to understand as you progress along your career. From a career development perspective and depending on how you would like to progress, having breadth might be important in the long term than a purely specialized depth. Depending on the company as well, many companies do value having a strong *depth* on a particular topic versus having general knowledge of many topics.

Advantages of breadth
- Being more malleable and understanding the interconnection of different processes.

- Being more open to change since specialization has not kicked in yet.
- *Less* short and medium job security than having a lot of depth; this is because generalists are *less* needed in times of low oil prices while generalists can thrive in boom or in times of economic expansion.

Advantages of Depth

- More critical in the short-term to medium-term for a particular project or organization.
- Being relied upon among your co-workers for expertise and advice in a particular subject.
- Ability to deeply understand a topic or area and give factual advice instead of having to rely on heuristics or other less reliable methods.

Absolute vs. Comparative Advantage

The concepts of absolute and comparative advantage are derived from economics and serve to represent why countries with different levels of development would engage in international trade. Comparative advantage is defined as the ability of a party to produce a particular good or service at a *lower marginal* and *opportunity* cost over another[82]. For example, Costa Rica has a comparative advantage in producing bananas while Saudi Arabia has a comparative advantage in producing oil, therefore Saudi Arabia can export oil to Costa Rica and import bananas from Costa Rica. Absolute advantage is defined as a party's ability to produce a certain good or service *more efficiently* than another party. Comparative advantage can be defined as a party's ability to produce a certain good or service at a *lower opportunity* cost than another party.

The same concept applies to individuals. Say employee A has an *absolute* advantage over employee B in *every* single aspect and assigned activity. Even though employee A can more efficiently complete *all tasks* required in a position and in a *shorter* period of time, it would still make sense to employ employee B, because that employee has a *lower opportunity* cost than employee A. For example, if employee A can make copies faster than employee B, it would still make sense to hire employee B since his or her *opportunity cost* of doing copies is lower. In other words, since employee A is doing copies he or she is giving up more other *higher value* items that they

---

[82] https://courses.lumenlearning.com/boundless-economics/chapter/introduction-to-international-trade/

could be doing *instead* of making copies. This is one of the reasons why tasks are delegated from higher and more productive employees to other less experienced employees.

Individual Contributor vs. Supervisory Assignments

There are two main types of roles in traditional human resources management in the oil & gas industry, an individual contributor and supervisory or managerial assignments. One of the early marks of a high potential employee in this industry is having a supervisor assignment early on in your career. Many of the employees that eventually make it to senior or executive management have had an early stint in a supervisory or managerial role.

Ability to delegate

Being able to delegate is one of the *key marks* of a successful supervisor and future leader. Think about any senior leader in the company and most if not all leaders have had long experience in delegating assignments.

Delegation is assigning responsibility and authority to someone in order to complete a clearly defined and agreed upon task, project or milestone while the supervisor or manager retains the *ultimate responsibility* for its success. Delegation incorporates empowering direct reports through effective leadership, and may be directed in any direction and used in any organization.

Why delegate in the first place? There are many benefits on delegating as a supervisor, among which are[83]:

- Efficiency: Delegation improves efficiency when it allows work to be transferred to people whose skills are a better match for the work. A supervisor is in charge of *planning* and *strategizing* the next steps for the team. When direct reports are able to carry out most of the routine activities required for the team, it allows the supervisor the time and effort needed to plan and strategize.
- Development: Supervisors often possess important skills and abilities that can be passed onto the direct reports. The best way of doing this is to coach them in the new skills and then delegate tasks to them so that they may use those new skills. Delegating is a great way of encouraging direct reports to develop themselves and for the supervisor to develop coaching and mentoring skills.

---

[83] https://projects.ncsu.edu/project/parkprgrd/PSTrainingModules/delegating/delsec1.htm

There are many reasons why supervisors may not delegate as much as they should, some of those beliefs are[84]:

- Believing that employees cannot do the job as well as the manager can.
- Thinking that it takes less time to do the work than it takes to delegate the responsibility.
- Having a lack of trust in employees' skills, motivation or attention to detail.
- Thinking that by not delegating we make ourselves indispensable.
- Enjoying doing the *actual work* more than delegating.

Accepting that not everything can be done by one person is the first step towards delegating. The important thing is not just to delegate, but to delegate *effectively*.

Ability to develop others to deliver results

One of the key responsibilities of supervisor positions is the ability to develop others, particularly the direct reports. A key mark of a successful future company leader is how often and how frequent the percent of their time is devoted to developing others and talent management. The higher an individual progresses throughout an organization the more important developing future talent becomes. In fact, key senior executives devote a significant portion of their time to *succession planning* and developing the next stage of company leaders.

Good managers attract candidates, drive performance, engagement and retention, and play a key role in maximizing employees' contribution to the firm[85]. The best managers ask, "How can we harness employee strengths, interests, and passions to create greater value for the firm?" Systematically linking organizational performance and individual development goals in the search for learning opportunities and better ways to work is a hallmark of organizations where sustainable careers flourish.

Vertical vs. Horizontal Moves

There are generally two types of career promotions, *vertical* and *horizontal* promotions. Vertical promotions or moves imply that an employee is moving to a higher position with usually a higher compensation, more

---

[84] https://www.shrm.org/resourcesandtools/hr-topics/organizational-and-employee-development/pages/delegateeffectively.aspx
[85] https://hbr.org/2014/01/if-youre-not-helping-people-develop-youre-not-management-material

responsibility and more impact to the overall company. A horizontal move is typically associated with going from one group, department to another or changing the focus of your current position, but retaining the same level of responsibility and compensation.

Vertical Move

Characteristics of a vertical move:

- Higher level of responsibilities than before.
- Generally, but not necessarily, higher compensation and benefits.
- If in a supervisory position, more direct reports than before.
- Higher budget and more visibility.
- May entail using the *same skills* or *different skills* depending on the type of promotion. For example, if an analyst is made a supervisor in the same group than before, the overall knowledge will be the same as before, just with a different focus now on *managing people* instead of being an individual contributor.
- If an employee has had a substantial vertical career growth throughout their career, they may lack the depth necessary for making decisions. This is particularly true if an individual has been promoted within the same function or department.
- Foster a *silo* mentality if vertical growth is only considered.
- May focus on a particularly set of functional or departmental level skillsets.

Horizontal Move

Characteristics of a horizontal move:

- Focus on learning a different group, business unit or area than before.
- Uses the same foundational skillset, but applied to *different problems* and offering *different* solutions. For example a mechanical engineer working in a refinery that is then subsequently transferred with the same title but working in an *upstream* business unit will learn a different type business.
- May involve transferring departments or doing a transfer within the same department but focusing on a different area. For example, a revenue accountant that handles one type of properties in Texas now handles revenue accounting for producing wells in Oklahoma.

- Overtime, may entail gaining more *depth* than *breadth* of knowledge, which for many *knowledge intensive* roles in the oil & gas industry is actually quite necessary.
- Gain a set of disparate skills, which increased marketability.
- Fosters a *cross-functional, cross-geographic*, matrix type thought instead of simply a hierarchical view seen in the same function vertical promotions.

Communication Skills

Communication skills are one of the most critical skills for an employee to have. Some experts even say that having good communication skills might be the most important skill in today's world[86]. One of the most valuable skills a person can have is to be able to communicate *highly complex* ideas in *simple terms* that people outside a particular field can understand. Think about great communicators like Steve Jobs, they were able to captivate audiences from around the world by having the gift of being a great orator and presenter. Steve Jobs was extremely talented in taking highly complex subjects, such as personal computers, smartphones or tablets, and being able to sell and market a product that would appeal to large percentages of the population. The following quote from Albert Einstein succinctly summarizes this concept:

> *"The definition of genius is taking the complex and making it simple."*

Communication skills are so important in the oil & gas industry, that many companies sponsor employees to take *Toastmasters* classes during the workday. Toastmasters is a nonprofit organization that operates clubs worldwide for the purpose of helping members improve their communication, public speaking and leadership. Toastmasters has currently more than 350,000 members and has more than 16,000 clubs around the world[87]. For new employees starting in the oil & gas industry, it is highly recommended that they become Toastmasters members and improve their communication and presentation skills. Toastmasters clubs generally meet over lunch at many workplaces around the country and the world. Toastmasters group can improve public speaking skills using a variety of topics.

---

[86] https://www.forbes.com/sites/gregsatell/2015/02/06/why-communication-is-todays-most-important-skill/#5434b4611100
[87] https://www.toastmasters.org/about/who-we-are

Additional communication requirements in the oil & gas industry

Due to the particular characteristics of the oil & gas industry, there are additional communication requirements to consider:

- As the nature of this industry is global, it is important to develop communication skills and strategies that can work with people from various countries or cultures. For example, in some countries, jokes are accepted as a way of creating rapport and camaraderie with coworkers but in other countries jokes have a negative connation and may imply disrespect. Knowing when to use or not to use jokes or other communication styles is an acquired and highly valuable skill. How to best communicate a complex idea to a wide and diverse audience is certainly a much demanded skill in this industry.
- As covered in this and the next chapter, the oil & gas industry is undergoing an *age* demographic shift not seen before. Many different age groups or *generations* work in different positions and teams, with *baby boomers* typically comprising upper management, with middle management being primarily comprised of *generation X*, it is important to tailor your style of communication to the audience.
- Too much communication inside a company or where companies restrict the use of communication due to the confidential nature of the information. For example, only a handful of individuals in a company often know the entire drilling program of a business unit or the company. There are many reasons why this is required, with confidentiality and competitor information being one of them, but this confidentiality restriction poses challenges to successful communication.
- Cross functional communication. A topic or subject that may be readily recognized by a function may not be for another. For example, a petroleum engineer may assume that everybody in the company knows about the different decline curves or the latest reservoir simulation software, or an accountant may assume that everybody is familiar with accrual accounting principles. Most of the time, and to facilitate cross functional understanding, it is best to define terms, that way a communicator can engage a higher proportion of the audience.

Successful communicator tips

- Know the topic very well; it is much easier to give a speech on a topic you are familiar and comfortable with than with other topics. Moreover, having gained the experience will increase the respect that the audience senses over you.
- Use a feedback loop; talk with audience instead of talking *to* the audience.
- Listen to nonverbal communication from the audience to adjust your speech.
- Be open to new ideas, encourage questions from the audience.
- Be specific and to the point.
- Be succinct; recognize that people's time is very valuable.

Nine-by-Nine Career Matrix

The nine box talent matrix is an easy to use tool that charts *potential* and *performance*. The following example provides a few of the categories used in the 9x9 matrix:

	Below Expectations	Meeting Expectations	Exceeding Expectations	
	Blocked or New Review role, train skills and provide mentoring by a Master.	**Future Star** Expand existing role and coach for imminent promotion.	**Super Star** Give stretch assignments and support with a mentor. Promote quickly.	**Rapid Progression**
	Under Performer Review cause of poor performance and train in role by a Master.	**Stalwart** Broaden experience within role and coach through any promotion.	**Current Star** Encourage in role development. Provide coaching prior to promotion.	**Cautious Progression**
	Misplaced Review role, their skills may be useful elsewhere. Otherwise exit.	**Professional** Encourage development within role and manage expectations.	**Master** Use to train others. Do not increase leadership responsibilities.	**Remain in Role**

Potential for Leadership Role →

Performance in Current Role →

Pareto's Law

Vilfredo Pareto, an Italian economist from the late 19th century, is credited with documenting what is commonly known as the "80-20" rule. Pareto started to notice that *20%* of the people in Italy owned *80%* of the land and this led to the study of this rule. The 80-20 rule states, that roughly 20% of something drives, owns, or is responsible for 80% of the results. In the context of career development, it is widely acknowledged that 80% of the results are generated by 20% of the employees. Or in other fields, 20% of the customers generate 80% of profits.

When establishing a new career path or position, the reader needs to understand and analyze the 80/20 rule in that particular position. For example, generally 20% of goals or results add 80% of the value. With the lower value tasks, the reader can automate these or delegate these tasks.

The same concept is applied in career management, which is a why many or most companies keep a list of what they call "HI-PO", which is short for High Potential.

High Potential (Hi-Po)

High potential employees are usually identified early in their careers, and are provided with *rapid advancement* opportunities or a fast track to management. High potential employees are usually those that can perform at 2-3 salary levels above their current state and have the *desire* and *willingness* to climb up the career ladder. It is important to note the distinction between *high potential* and *high performance*. All high potential employees are *high performance*, but not all high performance employees are *high potential*. This is an important distinction to make, since a high performance employee may not necessarily want to move or may not have the potential to get promoted.

Several scientific studies, in line with the 80/20 rule, have shown that, *across a wide range of tasks, industries, and organizations*, a small proportion of the workforce tends to drive a large percentage of the results[88]:

- the top 1% employees account for 10% of results
- the top 5% employees account for 25%, of results
- the top 20% employees account for 80% of results

Companies in the oil & gas industry, and particularly the integrated companies, tend to place a higher level of investment into high potential employees. These high potential employees are usually identified early in their careers and have specialized career development plans that would allow them to have a *breadth* of knowledge across the organization.

In a recent survey conducted by Harvard Business Review, over 98% of companies reported that they purposefully identify high potentials and that

---

[88] https://hbr.org/2017/10/what-science-says-about-identifying-high-potential-employees

they place a disproportionate attention on developing these employees deemed high potential[89]. Moreover, over 93% of companies who responded to the survey said that high potentials get promoted faster than other employees[90].

Qualities of High Potential Employees

High potential employees can exhibit several characteristics:

- Have demonstrated high performance in past assignments.
- Know the business well.
- Have aspirations for leadership opportunities.
- Have the potential to perform at several pay grades above their current levels.
- Work well autonomously.
- Exhibit leadership characteristics, such as ability to make decisions with limited or incomplete knowledge.
- Take initiative and are self-learners.
- Champion change.
- Are able to see the big picture while not getting bogged down in details.
- Have a willingness to innovate and take risks.
- Possess high *emotional* intelligence.
- Have an optimistic outlook at work.
- Have created an internal network of contacts.
- Exhibit the *right behaviors* consistent with company's expectations.
- Have the ability to shift from *individual or sole contributor* to *team leader* or manager role and inspire and lead employees to perform at the next level.
- Possess a good amount of *soft* skills, such as good communication, public speaking, and usually charismatic with the audience.
- Are adept at *change management* and can embrace change and are comfortable with not having all the details to make a decision.

---

[89] https://hbr.org/2010/06/are-you-a-high-potential
[90] Ibid

How do you know if you are considered high potential?

Most companies do not tell high potential employees that they are on the high potential list[91]. There are several ways to find out if you are considered a high potential employee:

- Do you get offered assignments that allow you to gain *breadth* and rotate at a *faster pace* than other employees?
- Are your assignments *challenging* and allow you to build skills *beyond* your core function? For example if you are an engineer and get offered an opportunity to lead a sales or finance team, then that typically means that the company has placed you on the high potential category since it is providing you with a *stretch* assignment.
- Do you get volunteered for high exposure and high impact projects, even though you may not necessarily have the experience? Then you also fall in this category.
- Does the company pay for you to pursue an advanced degree, such as an MBA from an Ivy League school or sends you to advanced training or seminars?
- Do you have an assigned mentor that is high up in the organization? When companies place their high level employees to mentor an early career employees that is usually a sign of being considered a high potential employee.
- Does the company place a high degree of *importance* on your career and educational development and rotate your assignments despite deadlines, requirements, or projects of your current role? Companies often move high potential employees *regardless* of upcoming deadlines, project completion or transition plans. If your company finds you *irreplaceable* or difficult to transition from your prior role, then you are generally not considered a high potential but more of a master or SME.
- What percentage of your time do you spend training others vs. the company training you? If you find a significant portion of your time is being spent training others and not receiving training or advanced conferences or seminars, then you are mostly not considered high potential.

---

[91] https://www.cebglobal.com/blogs/high-potential-employees-why-you-should-tell-them-theyre-hipos/

High Potentials vs. High Performers

Although the two tend to be mistaken, *potential* is different from *performance*. High performers are relatively easy to recognize and consistently exceed expectations, are management's *go to people* for difficult projects because of their successful track record. High performers are great at their roles but may not have the potential or desire to succeed in a higher-level role or do more advanced work[92].

Cross-Functional Assignments

Cross-functional assignments are often defined as those that take an individual from a particular function or discipline, such as engineering, finance, commercial and place that individual in a different function. For example, say John is currently a process engineer at a refinery, and leadership in that company sees that individual as having the potential to become a refinery manager or higher position. John could be moved from that process engineer role into a commercial role to gain business knowledge and commercial experience. This engineer would then gain more *broad* business or company skills that they would not get by becoming the best mechanical engineer in that refinery.

In these types of assignments a person can go from *knowing the most* to *learning the most*, which many individuals enjoy but others may not.

Some initial functions are more conducive to be moved into. Generally, roles that are more managerial in nature, can build on the skills an individual acquired from prior positions or assignments. Roles that require extensive detailed and specific knowledge, such as compliance, tax, safety or operations are typically not conducive for being moved into.

These types of assignments can be sometimes temporary or as time passes permanent. The intent of these assignments is to increase the breadth of knowledge of a high potential employees and improve that employee's skillset which will serve them well in future higher positions.

The fact that the company is willing to invest *time* and *effort* in developing that employee's career is a sure sign that the company's management sees significant potential for that employee to perform at several levels above his or her current status.

---

[92] https://www.softwareadvice.com/resources/high-potentials-vs-high-performers-a-managers-guide/

Stretch Assignments

A stretch assignment can vary from a specific task, position assignment, or project that is taken to develop an experience or expertise outside an employee's previous experience or core expertise. In fact, one of the key developmental tools to develop high potential employees is to provide them early on in their careers with as many stretch assignments as possible. Many high potential employees can point to that *key stretch assignment* that developed them on the job and prepared them for the position they have today. The more development the employee experiences in the stretch assignment, the more the company thinks that employee as a high potential. If you are offered a stretch assignment, even though you may not have the necessary skills in that area, it is always advisable to take on that assignment.

When assessing whether an assignment is considered a "stretch assignment" or not, it is important to understand what type of skillsets you will be gaining:

- Will be you gaining new skills that are predominantly outside of your comfort zone? For example if you are seen as an introvert in your day-to-day role, will this new assignment allow you to conduct presentations and become more of an *extrovert*?
- Will you be interacting with the same people that you work in your department or function, or will you be interacting with employees in other departments? The more interactions you have with employees outside of your immediate group, the more "stretch" the assignment is usually considered.
- If you are in a support function, will you be visible and interacting on a daily basis with business leaders or are you interacting within your support function? The more *critical* a project or assignment is, the higher it will impact the organization and thus the more input it will require from *decision makers* in the business unit.
- Does this stretch assignment involve increasing revenues or margins, thus impacting the company's bottom line? Or does this assignment involve reducing or managing costs? The more *impactful* assignments tend to have a *revenue* or *margin* improvement focus while those are that more *within* a function tend to have a *cost reduction* focus.
- Will you be primarily dealing or interacting with external people, such as customers, suppliers, investors, or government officials?

Usually, the more external-focus the assignment, the more it fits into the "stretch assignment" category.

When employees take on stretch assignments, they gain new skills, experiences and knowledge, allowing them to be more valuable to their employers[93]. There are many benefits to both the employer and employee of taking on a stretch assignment:

- Cost savings: by developing internal talent in a company instead of recruiting for outside hires, a company can reduce substantial hiring and training costs.
- More agile workforce: a company would have employees with much more diverse experiences and set of skills, leading to a more *adaptive* organization benefiting from cross pollination of knowledge.
- Higher employee retention: companies that encourage employees to learn skills and knowledge by providing stretch assignments, have significant higher retention rates have employees with more fulfilling work. When employees feel valued, they often work harder and become more engaged with the company.
- Faster development for the individual involved, with employers using those individuals who have participated in stretch assignments as future leaders of the company.

Types of stretch assignments

In the oil & gas industry, there are several types of stretch assignments that could be assigned to an individual:

- Department or functional stretch assignments, such as those that strengthen a candidate's *weaknesses*. For example, if employee A is seen as a high potential but does not have experience in public speaking, he or she may be rotated to a role that requires giving presentations to different audiences in order to improve that high potential's overall career future.
- Leading or being part of a team working on a large project, such as a new system implementation.
- Starting operations in a new country, state or city.
- Leading an entirely new business or downsizing an existing, non-performing one.

---

[93] http://www.gfoa.org/sites/default/files/GFR061624.pdf

- Rotating to a different location or business unit that is different from the current assignment.
- Business development with external counterparts.
- Learning a new tool or system.
- Address an upcoming market, regulatory or governmental change.

The Importance of having patience

Throughout your career you will have *ups* and *downs* related to your career progression. Patience, particularly in the oil & gas industry, is seen as a sign of maturity and showcases the ability to endure. Patience is also a measure of emotional intelligence and having the ability to deal with adversity and focus on the long-term.

> *"Patience is the companion of wisdom"* – Saint Augustine

> *"The key to everything is patience. You get the chicken by hatching the egg, not by smashing it"* – Arnold Glasow

> *"Patience is bitter, but its fruit is sweet."* — Aristotle

There are many benefits to a steady career progression in the oil & gas industry, such as:

- Companies provide incentives to reward employees to stay with the same company for several years, such as stock options that mature in three or more years, 401k vesting, pension plans that do not vest or have full value until 50 years or older.
- *Social capital* within the same company, such as having a network of colleagues that understand your work and your potential.
- You might be able to transfer to another department or function within the same company and still retain your service history, company knowledge of businesses, office politics, and many other advantages.
- Whether your current company is good or bad, you already have the knowledge of *what works* and *does not work*. In a new company the *unknown* could be a potential career *derailer*, such as company culture, supervisor assessment, ranking systems and many others. Sometimes the grass may *appear* greener when in *reality* it is *not*.
- For a significant percent of companies in the industry, the majority of executives and senior management will likely be long-term career employees. The only times companies hire external high

senior management employees is when they are undergoing major transformational change, restructuring, or dealing with bankruptcy.

- Patience is not complacency. Patience creates space to evaluate. Patience reframes situations. Patience has a bias for progress instead of perfection. Patience sees the war, not just the battle. It's gratitude for the journey, manifested as a persistent focus on growth, no matter how slow it may seem. Complacency is completely different: consistent meddling in mediocrity with no interest in moving forward. It confuses failure with finality, and paralyzes dreams in favor of practicality[94].

The following quote from Jeff Bezos, CEO and Founder of Amazon, explains why having a long-term approach has made Amazon great[95]:

If everything you do needs to work on a three-year time horizon, then you're competing against a lot of people. But if you're willing to invest on a seven-year time horizon, you're now competing against a fraction of those people…Just by lengthening the time horizon, you can engage in endeavors that you could never otherwise pursue.

Impatience can be dangerous

As in we live in modern times where *convenience* and *fast results* are the norm and *not* the exception, having impatience can actually be a *career derailer*. *Instant gratification* is not compatible with a long-term career, increasing wealth or personal happiness. It is important to understand that many things are beyond our control and this is very applicable in the oil & gas industry:

- Careers for most people would tend to advance when times are good, commodity prices or margins are high while career growth would tend to slow down when commodity prices are low. Since market conditions are beyond our control, it is important to understand the *ups* and *downs* of the business cycle and practice patience when possible.
- Mergers, acquisitions and divestitures, which depending on the individual's condition can be a career *accelerator* or a career *decelerator*.
- Offshoring, outsourcing and layoff events, particularly if the work being performed is in a support function.

---

[94] https://www.brazen.com/blog/archive/career-growth/slow-down-geny-why-you-should-practice-patience-in-your-career/
[95] https://hbr.org/2013/03/stop-fast-tracking-your-career

- We usually tend to get impatient *just* when things are about to change. For example, if an employee has been in a role for three years or so, there's a higher probability of having a new role in the next few months than it was at the beginning of the role.
- Emerging new projects, new corporate strategies or new management can impact our expected career plans in a positive or negative way, but developing a strong sense of patience and endurance can have far reaching impacts to a person's career.

We tend to think about careers as being a *straight line* of seamlessly going from point A to point B, but reality is significantly different than what we plan.

Below are examples of successful people that did not plan to become what they reached out. Their careers and challenges were like everybody else in these that no career path is a straight line.

- Jeff Bezos was working in one of the most reputable Wall Street investment banks in Manhattan in the 1990s and left his career to start an online bookstore in Seattle, WA.
- Henry Ford failed many times before finally succeeding with Ford Motor Company many years later.
- Greg Garland, current CEO of Phillips 66, did not want to move initially to his assignment in Qatar, an assignment that eventually led him to become CEO of ChevronPhillips Chemical and later on become CEO of Phillips 66.

Stanford marshmallow experiment

In the 1960's a Stanford University professor conducted an experiment with hundreds of children. The experiment consisted of having a small child come in to a private room and be seated in a chair with one marshmallow on the table. The adult conducting the experiment would offer the child a deal that if he or she would refrain from eating the marshmallow, they would be rewarded with a second marshmallow after the adult came back. For some children as soon as the adult exited the private room they would eat the marshmallow, other children would wait for a couple of minutes but the temptation proved too big so they would eat it. A few children would wait the entire 15 minutes to get rewarded with a second marshmallow.

The experiment results were published in 1972 and the children who participated in the experiment were studied later on. For the children who delayed eating[96] the first marshmallow they were found to have had a more successful life in many aspects[97]:

- Achieved higher SAT scores.
- Had lower levels of substance abuse.
- Had lower likelihood of obesity.
- Had better responses to stress.
- Had better social skills as reported by their parents.
- Generally achieved better scores in a range of other life measures.

What this experiment shows is the importance of patience and having the concept of *delayed gratification*[98], a concept that is critical in life as well as in career development.

Performance Management

The performance management process is instrumental in providing input to your career development plan as well impacting future assignments. Overall, goal setting for the year generally occurs in the first few months and performance reviews occur mid-year and towards the last months of the year. Many large companies use different types of performance management software such as *Success Factors* or *Taleo* or other applications, but the concepts are roughly the same.

Smart Goals

Most companies use some sort of variant of SMART goals[99]:

- Specific: what exactly do you want to achieve?
- Measurable: establish clear definitions to help you measure if you're reaching your goal.
- Attainable: what steps can you take to reach your goal?
- Relevant: how will meeting this goal help you or your company?
- Time-bound: how long will it take to reach your goal?

The goal setting process can vary not only among different companies, but also between different functions and departments. If for example an employee works in *shared services* organization where coworkers might do similar tasks, goal setting would be highly standardized and goals would be

---

[96] https://jamesclear.com/delayed-gratification
[97] https://jamesclear.com/delayed-gratification
[98] https://positivepsychologyprogram.com/delayed-gratification/
[99] https://emplo.com/blog/performance-management-best-practices

cascaded and almost equal. The same applies when employees work in the same department and have roughly the same level of experience and are located in the same salary bands or ranges.

Stretch Goals & Results

When setting goals it is always important to think in terms of *stretch* goals, or goals that would set an employee apart:

- Does this goal or result impact only my immediate department or does it impact the entire function or company?
- Compared to the goals and responsibilities of my peers, is this goal *above and beyond* of what's required in an employee typical of my salary band?
- Does this goal make me grow both *personally* and *professionally*?
- Is this goal based on my current skills or on developing new skills?

The importance of documentation

No matter the size of the company, having a document trail of results and achievements is very important. A best practice is to keep track of goals throughout the year in a spreadsheet or a Word document as to facility the mid-year and year-end review. Another good practice is to keep any positive feedback in Microsoft Outlook or similar email application and use it as reference during the final performance review.

Customer or Colleague Feedback

In many performance reviews processes it is very common to ask a group of *internal customers* or other company employees you interact on a regular basis to provide feedback on how you have performed during the year. Typically they may be asked a series of survey-like questions to ascertain your overall performance ranking.

Calibration & Salary Setting Process

After an employee has had their final review with their supervisor or manager, the supervisor needs to *calibrate* or rank his performance against other direct reports. This process is then repeated upwards and each individual is ranked for his or her performance based on their respective salary bands or pay scale groups. For example, how does this individual's results for the year compare against other individuals with the same salary level? Is it higher or lower? If it's higher then, the salary increase that comes as part of *cost of living* adjustments would be higher than peers. Overtime, having a high rating can impact not only future promotions but also overall compensation, including salary, bonuses or profit sharing as well as stock incentive plans.

Sample Salary Calculation

Similar to the effect of compound interest, having a higher rating can impact future salary increases. Let's assume an employee starts a position with a yearly salary of $50,000 per year, and he or she can get three types of ratings. The meet expectations category has a salary increase of 2%, while the mostly exceeds expectations gets assigned a 3% salary increase, while the always exceeds expectations get assigned a 5% salary increase. The table below illustrates the effect of compound salary increases, similar to the effect of compound interest when you save money at the bank for a long period of time, such as a 40-yr old career in the industry:

	Meets Expectation Rating	Mostly Exceeds Expectations	Always Exceeds Expectations
Typical Salary Increase Percent	2%	3%	5%
Year 0	$50,000	$50,000	$50,000
Year 5	$55,204	$57,964	$63,814
Year 10	$60,950	$67,196	$81,445
Year 20	$74,297	$90,306	$132,665
Year 30	$90,568	$121,363	$216,097
Year 40	$110,402	$163,102	$351,999
Lifetime Earnings	$3,130,501	$3,933,165	$6,391,988

As can be seen from the prior table, the effects within the first few years are not that much of a difference, but by year 30, the employee who consistently received the highest rating, would have a salary of more than $350,000 per year compared with the employee with an average rating who would get $110,000 per year. On a lifetime basis, assuming a 40-year career in the industry, the employee with the highest rating would receive $6.4MM versus $3.1MM. The other area that this calculation does not factor in is all the associated performance rewards, such as stock awards and bonuses/profit sharing, which add an even bigger *compounding* effect. An employee with the highest rating, depending as well on promotions and business needs, would be expected to make anywhere from two times to sometimes as much as five times as much over their lifetime than an employee who gets regular ratings. Keep in mind that many companies award a higher bonus percentage based on the employee's rating as well as stock options.

Career Development Plans

Career development plans, also known as Individual Development Plan (every company has a different acronym), is a document or series of documents that captures what your current state is and where you would like to be in your future state. This document varies from company to company, but commonly incorporates the following:

- Future possible positions or job titles that you would be interested in your career.
- Mobility preferences, whether somebody is willing to relocate within the home country or internationally.
- Assess skillsets, possible gaps and how to fill these gaps to achieve a future position.
- Develop a future state as far as what activities or roles would help you achieve your long-term goals.
- Define key milestones along the way.
- Provides a way for the management team that handles career promotions as a possible blueprint to follow.

The importance of being flexible

It is important that a career development plan is detailed and has long-term goals, but that it is *flexible* enough to accommodate unknown events along the road. Being flexible in your career goals is one of the most important lessons to learn while working and developing your career. Most people's careers are not a straight line of going from point A to B. There will be horizontal or lateral moves that help you develop necessary skills for your long-term goals. Usually, the roles that people think would not be of help are actually the roles that make an employee stretch and grow as a person. As Steve Jobs said in his very famous Stanford Commencement Speech:

> *"You can't connect the dots looking forward; you can only connect them looking backwards. So you have to trust that the dots sill somehow connect in your future. You have to trust in something – your gut, destiny, life, karma, whatever. Because believing that the dots will connect down the road will give you the confidence to follow your heart even when it leads you off the well-worn path; and that will make all the difference"*

Fast Lane Developmental Programs

The developmental of accelerated career development programs for high potentials has been around for several years. These programs generally combine the following features:

- A series of short assignments, which can last anywhere from 6 months to 18 months, and are designed to give rapid exposure to high potential employees to various parts of the business.
- Are run from a functional perspective, such as an engineering, finance, geoscience, procurement, operations, and other disciplines.
- Are used to develop a *pipeline* or *inventory* of future leaders for the company who are identified early in their careers so that their development can be shaped or influenced all throughout.
- Are given frequent access and exposure to senior company or functional leadership.
- Participants are identified early in their career or transitioning in the sector, such as MBA development programs.
- Participants are selected based on the *potential* to perform at several positions above their current position and compensation.

One of the industry's oldest management development programs, Chevron's Finance MBA Development Program, has been in place since 1946[100]. These development programs usually involve rapid rotation through several functional groups.

Sample development programs from other companies:

- Chevron's Horizons Development Program: which is a five-year training program designed for recent university graduates and combines assignments in various locations, as well as formal classroom instruction[101].
- ConocoPhillips/Phillips 66 Finance Excellence Program: which is a 3-6 month rotational program where participants learn about the company's operations along the entire value chain, as well the impact of those operations in the financial statements and performance metrics. Participants are exposed to key industry issues, business drivers, and company strategy[102].

---

[100] http://careers.chevron.com/students-and-graduates/graduate-programs/finance-MBA-program
[101] http://careers.chevron.com/professionals/programs#horizons development program
[102] http://careers.conocophillips.com/career-areas/finance-careers/

- Shell Graduate Programme: which primarily focuses on Commercial or Technical functions, is a 3-5 year program where students apply before graduating university[103]. Typically participants have up to three assignments for the duration of the program and can work in local and global locations. The program provides a combination of formal training in their particular business unit or function (upstream, downstream, commercial and others), formal checkpoints, mentoring and coaching as well as project assignments along each rotation. Shell's program is one of the most comprehensive programs and also one of the few in the industry where students can apply directly *before* starting with the company.

Best practices in Talent Management

There are several best practices in talent management that are worth mentioning:

- Identify and reward high performing employees.
- Conduct regular employee reviews to keep employee performance on track.
- Provide on-going feedback in areas of strengths, weaknesses, and future opportunities.
- Provide cross-functional assignments, particularly to those identified as high potential employees.
- Monitoring talent and measuring how talent management is performed with key indicators.
- Provide assignments and opportunities to develop new skills such as with formal training opportunities, mentorship opportunities, or new project assignments.
- Be forward thinking and start developing *bench strength* for succession planning. Bench strength is the concept of how many capable or ready employees a company has to fill out key positions in the company in case of retirements or employees leaving the company.

---

[103] https://www.shell.com/promos/careers/shell-graduate-programme-interactive-pdf/_jcr_content.stream/1488887214861/00760f90514efe706a1f0c44925ea0c0d88e22157853150259b942367cdc9862/shell-graduate-programme-guide-2017.pdf

From a company perspective, there are many benefits of having a great talent management system. In general, companies that align talent strategy with business strategy can experience the following benefits[104]:

- Boost in employee morale.
- Increased productivity.
- Lower turnover in employees.
- Increased talent capability and ability to grow internally. This is particularly critical in the oil & gas industry, where a significant portion of higher level employees are expected to retire within the next five to ten years.

Public Speaking Skills

Public speaking is one of the most valuable and critical skills any individual can have. From times immemorial the ability to effectively and succinctly convey ideas and convince people has been highly sought after, especially in the oil & gas industry. If you look at the senior executives of most publicly traded energy companies, you'll find out that *most* if not *all of them* have excellent to extraordinary public speaking skills. The ability to tell a story and effectively communicate is one of the skillsets that separates *followers* from *leaders*.

In fact one of the required skills to climb up the corporate ladder, particularly coming from a technical background, is having adequate public speaking skills. Managers must be able to clearly communicate goals, requests and other messages to their teams, as well as be able to speak at the *right level of detail* based on the expected audience.

Here are some of the benefits associated with public speaking:

- Makes your job easier.
- Promotes your reputation & confidence about a topic[105].
- Improves your career options, by being able to participate as an individual contributor as well in a supervisory or managerial role.
- Ability to easily assume leadership in a group or department[106].
- Develops your *listening, reading,* and *writing* skills by helping you prepare for presentations and other venues or occasions where you can use prepared remarks.

---

[104] Korn Ferry Institute, Talent Management Best Practice Series, Strategic Alignment
[105] https://engineeringcareercoach.com/benefits-of-public-speaking/
[106] https://www.write-out-loud.com/benefits-of-public-speaking.html

- Allows you to share knowledge, motivate others and help people help themselves[107].
- Speaking well about a subject can increase credibility, confidence, and perception in the organization.
- You increase your emotional intelligence by controlling your thoughts and emotions.

---

[107] Ibid

Chapter IV – The War for Talent in the Oil & Gas Industry

"If one does not know to which port one is sailing, no wind is favorable." – Lucius Annaeus Seneca

The importance of choosing the right career

We spend a significant time of our lives at work. The average worker typically spends 2,000 or more hours per year, meaning that over the average working career span of 40 years a person would spend 80,000 hours or more at work! It is therefore quite important to choose wisely and have a career that fits your current needs and long-term goals. There are many long-term impacts of choosing the right career:

- Satisfaction
- Compensation
- Engagement
- Healthier life
- Personal development

Choosing the right career is one of the most difficult decisions any person can do in his lifetime. This is one of the key questions as to why we decided to write this book, to help readers with more information make a better career decision.

Asking the Right Questions

The following are sample questions than can be used to assess what a future career would look like[108]:

- What are you really good at?
- What do you enjoy doing?
- What is your appetite for work/life balance? Do you expect it to change in the future with certain life events (i.e. getting married, having children, caring for older parents, and others)?
- What do you want to avoid in your career?
- Do you like repetition? Do you like stability?
- Do you enjoy having a new challenge everyday or prefer to have a more structured role?
- What experiences or skills must you have in order to succeed in that career?
- What does your ideal career or job entail?
- Are you introverted or extroverted?

---

[108] Some inspiration taken from: https://www.comparebusinessproducts.com/briefs/15-most-important-career-questions-ask-yourself

- Do you enjoy working with people or more with systems or processes?
- Do you like to do public speaking and presentations?
- How important is it for you to have a top of the line compensation versus being fully engaged and passionate about your work? The two may sometimes be mutually exclusive, with many highly compensated employees reporting less engagement at work than those with more meaningful or impactful careers in other aspects (i.e. many people enjoy coaching, mentoring, and training others even though the compensation or corporate-wide impact may not be as high as other roles.)
- Do you like to have a planned day & week versus having new challenges? Do you like to be in "firefighting" mode constantly? Many functions in the oil & gas have more predictability than others, with those in certain support functions have more of a *predictable schedule* (i.e. Finance with their workday task-driven calendar vs. a trading function where every day there might be a new geopolitical or unexpected event that impacts the markets.)

Maslow's Hierarchy of Needs

American Psychologist Abraham Maslow, developed the theory of self-actualization[109]. In this theory, he identified the five groups or hierarchy of needs[110]:

- Biology and physiological needs, such as air, food, drink, shelter, warmth, and sleep.
- Safety needs, which is protection from elements, security, order, stability, and freedom from fear.
- Love and belonging needs, friendship, intimacy, trust and acceptance and receiving and giving affection. Being part of a group (family, friend, work, hobbies).
- Esteem needs, which often involve esteem for oneself such as having dignity, achievement, mastery and independence as well as the desire for reputation or respect from others.
- Self-actualization needs, realizing personal potential, self-fulfillment, seeking personal growth and peak experiences.

---

[109] http://blog.degreed.com/maslows-hierarchy-of-needs-for-your-career/
[110] https://www.simplypsychology.org/maslow.html

The following table provides a summary of how this is applicable at work and choosing a career[111]:

Physiological needs	Having a clean working space, supplies, equipment, technology and overall physical comfort at work
Safety Needs	Reasonable and predictable income, benefits, lack of culture of fear, safe work environment
Love and Belonging Needs	Feeling support at work, sense of belonging at work. Teamwork, mentorship and sense of acceptance as well as feeling that you are making a contribution towards end goals
Esteem Needs	Feeling that your work matters, is recognized by others, mastering concepts and becoming an expert builds up your esteem
Self-Actualization Needs	Realizing your full potential as individual, learning how and where you can apply your knowledge that can result in peak experiences that make you a better employee

Throughout your career your needs might vary or you might have different needs at different points. For example, earlier in your career and depending on your personality, as long as job provides for physiological and safety needs you might be fully satisfied. As you advance more and more in your career, the other three needs start manifesting with the ultimate one, self-actualization, being the most difficult to attain generally.

The important concept is to assess where you are in your current role, determine which needs are not being met, typically the latter two, and understand the different options as far satisfying those. Do you need to change companies to achieve the last one? Or would a rotation in a different function or business unit *within the same company* allow you to achieve those?

How needs progress throughout your career

These hierarchy needs might fluctuate as you rotate from position to position. Say for example that you are considered a subject matter expert in a particularly defined area. In that current role, all of your hierarchy needs may be met, including self-actualization as you are seen as the go-to person and can readily and quickly help the company and employees solve problems. Let's say you get promoted and you rotate into a new position, for the first few years the last two needs may not be fully met for some time. One of the key transitions in growing your career comes from stretching your knowledge, but that makes the individual vulnerable in a sense while that individual gains knowledge to get more comfortable. Many SME's experience this when going from one area where they are seen as the

---

[111] http://blog.degreed.com/maslows-hierarchy-of-needs-for-your-career/

go-to person to another area where they are a complete newbie. Being comfortable in not knowing everything about a new position can help cushion this change of going from being the smartest person in the group, who can contribute, has authority on the subject, and whose opinion is sought after to being a follower for some time in an entirely new position. The following table illustrates some of the ups and downs people many experience from a Maslow's hierarchy of needs as they rotate through new positions:

Year	Position	Description	Physiological needs	Safety Needs	Love and Belonging Needs	Esteem Needs
5	Position 1	Employee is the "go-to person" and is highly praised by everybody	Met	Met	Met	Met
6	Position 2	Employee is rotated to a new group and struggles for the first few months	Met	Met	Not Met	Not Met
10	Position 3	Employee is rotated to a similar position and achieves proficiency and praise fairly quickly	Met	Met	Met	Met
10	Position 4	Employee is rotated again and cycle starts	Met	Met	Not Met	Not Met

Depending on how the transition into the new position goes, even the safety needs might be challenged. For example, a highly esteemed SME might feel comfortable in their career stability and in the fact his or her services are highly sought after. When that person moves from that role, the safety needs may not be met or might be challenged for some period as that person may feel less valuable internally as their short-term contributions diminish.

Why work for an Oil & Gas company?

If you are starting your career in the oil & gas industry, it is very important to know and realize why you may decide to go in this industry and why work for an oil & gas company. There are many benefits of working for an oil & gas company, among which are:

- Vital industry to the world, people will always need energy in one form or another.
- Complex industry with ever changing regulations, requirements, and challenges.
- Geographically diverse industry with many locations around the world.
- Career opportunities in many fields and functions, from engineering, to commercial, to accounting, to geosciences, a diverse set of skills are necessary to safely produce hydrocarbons.
- Higher compensation & benefits than most other industries.
- An industry with a high emphasis on career development, mentoring opportunities.
- Highly capitalized or funded pension benefits, with many oil & gas companies offering defined-benefit programs, as well as extensive 401K matching.

How to search for a job in the first place

Although there has been growth in the online application processes and there are several application systems, the best way to find the next position is through networking. Recent studies have shown that over 70% of all positions are filled through referrals[112], with some studies showing that less than 15% of positions are filled through online applications only[113]. Even more challenging is the fact that almost 80% of all new jobs are not even listed in the first place.

Networking

Networking is one of the most critical aspects of searching for a job and you should actively network *before* needing to search for a job. Very few positions are actually filled through the application tracking systems described in subsequent pages. Networking is generally defined as *"the cultivation of productive relationships for employment or business."*

---

[112] https://www.payscale.com/career-news/2017/04/many-jobs-found-networking
[113] https://www.linkedin.com/pulse/new-survey-reveals-85-all-jobs-filled-via-networking-lou-adler

Networking is one of the most if not *the most* effective job searching strategies you can do. Networking is not just attending an event or using LinkedIn. According to the CEO and founder of the Adler Group "networking is how you turn 4-5 great contacts into 50-60 connections in 2-3 weeks… networking should represent about 60% of your job-hunting efforts or about 20-30 hours per week[114]." Another important factor is that by networking you increase the probability of somebody who you did not expect that could be able to assist. Besides, it is always a good practice to cultivate the contacts in your network on a frequent basis so that when you do need to have them help you; your memory is fresher in their mind instead of being somebody who they have not contacted in 10 years.

Recruiters

Several companies make extensive use of recruiters or recruiting firms. Many of the large companies can sometimes pay up to 20-40% of a new employee's salary for finding the right candidate. The compensation they will get also depends on the level of the position being filled, with executive positions sometimes being compensated 1-3 times salary of that employee.

Companies in the oil & gas industry use recruiters since they can perform the following tasks:

- Committed to their clients since they have an incentive to continue to provide quality candidates, and many times providing a guarantee if the candidate leaves or does not perform within a pre-established timeframe, the recruiter will provide another candidate free of commission.
- Recruiters are more in tune with the job market and may work for several companies, thus have a large list of possible candidates with a wide variety of skills and specializations.
- Recruiters have direct access to human resources or hiring managers and have a vested interest in having the right candidate having an interview with the company.

However, there are certain drawbacks that you should be aware when working with recruiters:

- Recruiters sometimes are required to bring several candidates to interview, with many times having a great candidate or the candidate they feel has the higher probability of getting that

---

[114] http://www.worknetdupage.org/blog/2016/01/20/networking-important-job-search/

position alongside with several *filler candidates* so that the comparison between them is obvious. Keep in mind that many candidates are brought over to interviews without any intention of actually hiring them, so you need to be aware and factor this in your job search.
- Recruiters have exclusivity requirements, so that they will represent a candidate not only for a *specific position* but also *actively discourage* to apply or contact the client directly. This requirement needs to be factored in as well.

Staffing Agencies

Staffing agencies are companies that have regular employee or contractors readily available to hire on projects or as on-going contractors to companies. Staffing agencies are sometimes called "contingent workforce" in that they supply staff to companies with a variety of skills and requirements for many functions in the oil & gas industry, including accounting, IT, engineering, operations, and many others. Some of the major staffing agencies include Adecco and Burnett Specialists. Keep in mind that there are many or even thousands of staffing agencies across the country and many specialize in different functions. Generally, staffing agencies are paid on a cost basis plus a markup or expected profit per hour for each staff member that they bill to a major company. For example, a staff member, who might be a W-2 employee of that staffing agency could be paid $40 per hour by the staffing agency, and then billed to an E&P company at $70 per hour, the difference being accounted by burden & benefit costs (health care, social security taxes, unemployment taxes, administrative expenses) plus a profit or gross margin.

Applicant Tracking Systems

There are a few positions in between that are truly posted online and allow you to apply for those. Often a job posting receives tens, hundreds if not thousands of applicants, making the typical online search to have a low probability of success. Why do employers use applicant tracking systems in the first place? These systems help employers save time; paper and help employers stay organized throughout the recruiting process and keep track of all relevant information regarding an employment candidate[115]. They were usually first started to be used by large corporations, but now more and more mid-size and smaller companies are using these ATS.

---

[115] https://www.jobscan.co/blog/8-things-you-need-to-know-about-applicant-tracking-systems/

Each time a candidate uploads his or her resume to an ATS, an entry in the database is created. The ATS "reads" the formatted resume in Word, PDF, or other format and catalogs this data. When the recruiter or hiring manager logs in into the system, what they see is not your submitted resume, but the "scanned" copy which may or may not include all the information you submitted. Applicant tracking systems contain different database fields for information on a resume, such as the candidate's name, contact details, work experience, job titles, education, employer names, and periods of employment. These systems try to identify this information on a job seeker's resume, but if a resume isn't formatted according to the applicant tracking system, it won't pull this information into the proper fields[116].

How do these ATS work?

If you so desire to search online, it is important that your resume conforms to the following guidelines:

- Each resume should be unique for each job positing or application, highlighting how your *experiences* and *skills* fit with this position.
- A resume should be optimized with relevant keywords associated with the posting. Applicant tracking systems rank each resume against each job posting based on the percent match of how it fits with the job description.
- Have an active LinkedIn profile that matches your resume.
- The resume should be formatted so that the machine can read it and parse the data accordingly.
- How unique or rare the keywords for that job are versus the resume.
- Never send your resume in a PDF format, since these ATS do not have a standard way of reading or structuring PDF documents[117].
- Do not include tables or graphics and they often misread tables.
- Length of resume does not matter, submit a longer resume that details your experience and increases the number of keywords, phrases and relevant experience, which increase the probability of your resume being seen and ranked higher.
- Call your work experience "Work experience" as these ATS will typically skip other titles such as "professional experience" and others[118].

---

[116] https://www.cio.com/article/2398753/careers-staffing/careers-staffing-5-insider-secrets-for-beating-applicant-tracking-systems.html?page=2
[117] Ibid

- Do not start your work experience with dates, always start with employer's name, your position title, and then dates you held that title. ATS look first for company names[119].

Common Applicant Tracking Systems

Most if not all applicant tracking systems have the same goals and functionality:

- Track applicants and create job requisitions for job postings.
- Market the company's available positions.
- Provide compliance with EEO and other government requirements.
- Create a database or talent pool available for future hiring.
- Sort out or "rank" applications using algorithms that read the candidate's resume and compares it against the job posting.

There are more than 315 Applicant Tracking Systems[120] used by various companies, from small to Fortune 500 companies.

The table on the next page provides a comparison of the different ATS software used by major oil & gas companies.

---

[118] Ibid
[119] Ibid
[120] For more information, please visit: https://www.softwareadvice.com/hr/applicant-tracking-software-comparison/

Sector	Company	ATS	ATS Parent Company
IOC	ExxonMobil	SuccessFactors	SAP
IOC	RoyalDutchShell	BrassRing	IBM
IOC	BP	BrassRing	IBM
IOC	Total	BrassRing	IBM
IOC	Chevron	ADP	SAP
Independent E&P	ConocoPhillips	Taleo	Oracle
Independent E&P	Occidental Petroleum	Taleo	Oracle
Independent E&P	Anadarko	Taleo	Oracle
Independent E&P	Concho Resources	Workday	Workday
Independent E&P	Pioneer Natural Resources	Workday	Workday
Independent E&P	Hess	SuccessFactors	SAP
Independent E&P	Continental Resources	Jobvite	Jobvite
Downstream	Phillips 66	SuccessFactors	SAP
Downstream	Andeavor (formerly TSO)	SuccessFactors	SAP
Downstream	Valero	Taleo	Oracle
Downstream	PBF Energy	Workday	Workday
Downstream	Delek Refining	SuccessFactors	SAP
Midstream	Enterprise Products	Taleo	Oracle
Midstream	Kinder Morgan	PeopleClick	PeopleFluent
Midstream	Targa Resources	Ceridian	Ceridian
Midstream	Plains All American	Ultipro	Ultimate Software
Midstream	Energy Transfer	SilkRoad	SilkRoad

Job Search Tips

If a job posting online has a very short duration from say posting date to closing, it is highly likely that the company already has a candidate already selected and it is just completing a posting for *legal* & *compliance* reasons.

Another sure sign that a posting is not intended for external applicants is a very detailed and *possibly contradicting* list of requirements regarding a position. For example, if an external posting says "5 years minimum experience with XYZ company processes" or very specific custom systems that an external hire would have not experience in specifically.

As mentioned before, one of the most effective ways to get an interview for a position is through a referral. There are also many other ways of getting a job interview:

- Searching through sites that aggregate job postings, such as the rigzone.com, LinkedIn, and many others.
- Since the majority of most job applications are received through ATS or online, another way to stand out is to use physical mail,

such as postal service, UPS or FedEx. One way to do this is to research the hiring manager through LinkedIn or other social media. Then with this contact information, address your resume, cover letter, and possibly sample of work to that hiring manager using the company's physical address.

Application Process Tips

There are several ways to improve the chance of having your application or resume be selected:

- Be sure to provide a cover letter that is *tailored* for each position. A cover letter should be seen as a "sale pitch" in a sense. You should use this letter to sell the company on *why* you should be hired for this position.
- Always use a professional, easy to spell and not too long email address. For a few dollars a year you can customize or buy a domain that allows interviewers to remember your name. Which one is more impactful? An email that says jks-1999@gmail.com or john@smith.com ?
- Use standard fonts for all of your electronic documents being submitted. As companies use ATS to screen out resumes, it is important to have ATS readable documents so that it can feed the company's recruiting database and rank you as a high fit candidate.

Networking

Networking is a critical portion of any job search, but it is particularly applicable in the vast oil & gas industry. This industry is characterized by having a large number of different organizations, such as the American Petroleum Institute, chambers of commerce, marketing and retailing associations, independent petroleum producers association, and many others.

There are several organizations within the oil & gas industry that promote networking among its members. The following are associations around oil & gas industry that you could join:

- Society of Petroleum Engineers is the largest association in the oil & gas industry. The SPE currently has more than 164,000 members in 143 countries with more than 198 sections or chapters around the world[121]. The SPE was originally founded in 1957 as a separate organization from the American Society of Mining Engineers and has grown from less than 20,000 members in the 1960's to more than 164,000 members currently. The SPE has many functions in the industry, and one of the most critical areas is in the area is financial recognition of petroleum reserves, called Guidelines for the Evaluation of Petroleum Reserves and Resources[122], which the SEC uses as a basis for basing their own reporting requirements. SPE members can become members at varying member rates starting at $20 per year to more than $150 per year[123].

- Young Professionals in Energy or YPE, which aims to facilitate the advancement of young professionals in the energy industry around the world through social, educational and civic service oriented events[124]. YPE has local chapters around the world, with a big concentration in energy-heavy cites such as Dallas, Denver, Houston, Calgary, Fort Worth, Oklahoma City, and many others. YPE has an extensive job listing section which members and non-members can use to look for a position in companies that sponsor the organization.

- Petroleum Accounting Societies are organizations that band together accountants and other finance groups who work in the industry, particularly in the upstream side. COPAS is the governing council of these petroleum accountant societies and is headquartered in Denver, CO. More information can be found on their website at www.copas.org. Also keep in mind that the many local societies have local job postings available to members in those local chapters. Currently there are more than 24 chapters in the United States.

Interview Process

Depending on the type of company, the interview process would tend to be highly structured and process-driven. For example, several of the majors only recruit on campus during the Fall semester in the United States and do not consider new hires out of college outside of this timeframe. The smaller

---

[121] http://www.spe.org/about/
[122] For more information, please visit: http://www.spe.org/industry/docs/GuidelinesEvaluationReservesResources_2001.pdf
[123] http://www.spe.org/join/dues.php
[124] http://ypenergy.org/

companies would tend to have a less structured approach on conducting interviews.

Different Types of Interviews

There are many different types of interviews, from telephone interviews, to in-person, one-on-one interviews, to group interviews to day-long assessment days[125].

- Telephone Interview
- Video Interview
- Panel Interview
- Assessment Day
- Group Interviews
- Individual (face-to-face) interviews

The table below and on the next page summarizes the different types of interviews, their advantages and disadvantages:

Type	Description	Advantages	Disadvantages
Telephone Interview	Telephone interviews speed up the candidate selection process and are usually conducted initially to screen out candidates and determine the level of interest. Typically phone interviews last up to 30 minutes	Quicker than a face-to-face interview Costs less to both interviewer and interviewee The employer can test the candidate's telephone skills Both the company and employee can screen out quicker Easier for the candidate to participate in the interview without taking time off from the current job	Signal would be weaker or call quality could be No way to read body language of either the interviewer or interviewee Difficult to establish rapport
Video Interview	Substitute of a phone interview. Video interviews can be setup through software such as Skype, Cisco, or other as well as iPhone's facetime or other technology. Often video interviews last up to 30 minutes	Quicker than a face-to-face interview Costs less to both interviewer and interviewee The employer can test the candidate's telephone skills Both the company and employee can screen out quicker Easier for the candidate to participate in the interview without taking time off from the current job	More prone to technological glitches or issues than a phone interview Have to be in a formal setting in order to be recorded correctly Depending on the video quality, interviewers can read body language Difficult to build rapport in this type of interview
Panel Interview	Similar to face-to-face interviews by multiple people. Typically can last from 30 minutes	A reduction of personal biases and probability of gaining rapport with different people All applicable people may	Depending on the personality, number of interviewers and overall mood, could be overwhelming for interviewee.

---

[125] https://www.coburgbanks.co.uk/blog/assessing-applicants/6-different-types-of-interview/

Type	Description	Advantages	Disadvantages
	to 1 hour. Can involve multiple interviewees with different styles	meet at the same time (i.e. for example from the same workgroup)	Risk of disagreement among the panel (good cop / bad cop)
Assessment Day	Used to interview a large group of applicants and is very commonly found among university graduates. That way the employer can assess among many options	Provides additional time to get to know each candidate Employer can interview all candidates at the same time, minimizing scheduling issues with interviewers Generates competition among the interviewees and employers can assess how candidate interact with each other	Dominant personalities/type A personalities will tend to dominate other candidates with different personalities. More difficult to build rapport with individuals Certain positions are not a good fit for an assessment, usually those that require *introvert* skills Requires interviewees to be available for a longer period of time in order to provide consistency
Group Interviews	Similar to the assessment day interview	Good fit for positions that require team work and soft skills over hard skills	Interviewers can play different roles (bad cop good cop) One interviewer would be focused on your reaction versus another one asking question
Individual (face-to-face) interviews	One interview per candidate. Frequently the hiring manager conducts the interview and makes the decision regarding the position. Commonly 30 minutes to 1 hour	Uses body language to communicate interest, rapport and other cues Candidate can see how his or her future supervisor will treat them and establish a familiar face The employer can test how the candidate perform under stress or difficult social situations	Time commitment and logistical issues for both candidate and interviewer. Interviewers can fall for charismatic candidates, not necessarily the most appropriate for the job Personal bias can cloud judgment of both candidate and interviewer
Video Interview	The interview is conducted through some sort of program that captures video or is recorded for later playback	Candidates can interview from the comfort of their homes More time to practice Best use and most flexible for candidate and interviewer	Technology always does not work as intended Problems with delays or lags Personal communication is best

How to prepare for an interview

Being prepared for an interview can be the difference between getting that dream job and not. The following are tips that job seekers might find useful.

Before the interview

- Research everything you can about the company and the person interviewing you. (Including social media, stock information, founder's bio, company's 10K, 10Q, etc.….)

- Read through your resume enough times to know it really well. Be sure to have key talking points or summary of how your experiences and skills fit in well with the position being applied.
- If you got fired, say you got fired and explain why. Take ownership of the issue. Odds are it was your fault to some degree and even if it was not the interviewer is not likely to believe it was the employer's fault.
- Prepare a list of questions to ask the interviewer. The section in a few pages covers sample questions to ask in a job interview.
- If you are leaving a current job, do not bad mouth your boss. Remember, the person interviewing you could be your boss. When you talk bad about your current or even old boss they will assume you are going to say the same about them.
- Decide how much information you are going to reveal about yourself. There are certain questions they cannot ask you. It is up to you to decide if you want to divulge that information.
- Go to the bathroom before the interview.
- If water or a drink is offered, take it. You might find yourself with a dry mouth in a few minutes.
- Ask if you can follow up with them in a certain amount of time, this shows that you are interested in the position and that you would like to proceed forward.
- Mail them a follow up letter or email as soon as you walk out the door so it is on their desk the next day or two.
- If they give you their business card, take it, look at it and store it in your pad folio. Email them within the same day thanking them for the time and ask any follow up questions.

Sample list of questions for a job interview

The following lists of questions are not meant to be comprehensive questions but just meant to be used as a brainstorming tool. Remember that the best way to impress your interviewers is to customize the questions for your particular sector in oil & gas as well as to have general knowledge on the company.

General position questions

- What distinguishes *great* employees from simply *good* employees?
- What are some of the typical responsibilities associated with this position?

- Are the responsibilities cyclical in nature (month-end close for example)?
- What are some of the biggest opportunities you see in your department or group?
- What are some of the expectations for a new employee coming in into this group for say the first month, 90 days, and 6 months?
- How critical are communication skills for this position?
- What type of personality works best for your organization?
- How open is the company to *failure*?
- Does the company use failure in a positive way to learn and increase opportunities for success in the future?
- Looking back at your career, what is your biggest accomplishment?
- Define what does *excellence* in this position entail?
- Looking back at prior successful and not so successful employees you have hired in the past, can you describe what those key traits were?

Position specific questions

- What application or software do you use on a day to day basis?
- What are some of the unique aspects of this group or functions?
- What are some of the biggest challenges employees in your group or department face?
- What is the typical career ladder for new employees?
- What are the scheduled job rotations for this position?
- How long have you been with the company and in your current group?
- What are the particular skillsets that you are looking for in a new employee?
- What happened to the person who was in the position before? Did they move to another group or did they leave the company?
- Are there a lot of contractor positions within your department?

Recent Event or Industry Questions

Whenever you are interested in a particular company, it is best to be informed and know about that company as much as possible.

- I listened to the company's recent earnings conference call and noticed that production volumes have decreased compared to prior years. How do you expect the company to mitigate these risks?
- What differentiates this company's strategy from other companies?
- Does the company tend to be a *trend-setter* or *trend follower*?
- Does the company's management tend to be recruited extensively from within the ranks or is it hired externally?

Preparation for behavioral interviews

Behavioral interviews are widely used by all kinds of companies in many industries. Employers use this type of interview to get insights into *how you* handle specific situations in the workplace. The interviewer will want examples of what happened in a particular challenging circumstance, what you did, and how you achieved a positive outcome[126].

The STAR method or Situation, Task, Action, Result, is a widely used interview method, particularly by large companies.

- Situation: Describe the situation that you were in or the task that you needed to accomplish. You must describe a specific event or situation, not a generalized description of what you have done in the past. Be sure to give enough detail for the interviewer to understand. This situation can be from a previous job, from a volunteer experience, or any relevant event.
- Task: What goal were you working toward?
- Action: Describe the actions you took to address the situation with an appropriate amount of detail and keep the focus on YOU. What specific steps did you take and what was your particular contribution? Be careful that you don't describe what the team or group did when talking about a project, but what you actually did. Use the word "I," not "we" when describing actions.
- Result: Describe the outcome of your actions and don't be shy about taking credit for your behavior. What happened? How did the event end? What did you accomplish? What did you learn? Make sure your answer contains multiple positive results[127].

---

[126] https://www.thebalance.com/behavioral-interview-techniques-and-strategies-2059621
[127] https://www.vawizard.org/wiz-pdf/STAR_Method_Interviews.pdf

Chapter V – Accounting & Finance

"Accounting is the language of practical business life. It was a very useful thing to deliver to civilization" – Charlie Munger

What is Accounting?

Accounting is generally defined as the systematic and comprehensive recording of financial transactions pertaining to a business as well as the process of summarizing, analyzing and reporting these transactions in financial statements[128].

Accounting has been a critical field in the oil & gas industry ever since the first *commercial* oil well was drilled in Pennsylvania in the 1800's. In fact, John D. Rockefeller always had an appreciation for accounting and was always financially savvy, which led to establish the largest oil company the world has ever seen, Standard Oil.

What roles do accounting & finance play in the oil & gas industry?

Accounting & Finance are a key support function in the exploration, drilling, production, transportation, processing and distribution of hydrocarbons. Some typical functions or activities performed by this function:

- Process revenues and payments to working and royalty interest owners and ensure these are correct.
- File production reports with the EIA, state and local agencies as well pay royalties, severance, and other taxes.
- Collect and apply cash from customers and ensure accounts receivable are current.
- Process invoices from vendors and pay these vendors on a timely manner, maximizing discounts, and handle vendor inquiries.
- Administer and disburse payroll to a company's employees, process withholdings and submit payroll and other employee taxes to the IRS and other government agencies.
- Reconcile customer, vendor, general ledger, and other accounts to ensure quality financial statements and reporting.
- Audit the company's processes and procedures to ensure compliance with operational regulations as well as SOX regulations on financial reporting and that the company's internal controls are being followed.
- Process business transactions from a variety of businesses and sectors within an oil & gas company, such as exploration,

---

[128] https://www.investopedia.com/terms/a/accounting.asp

production, transportation, refining, marketing, and fractionation of hydrocarbons.
- Analyze the company's financial reports, research, and explain variances versus prior periods as well as versus plans & budgets.
- Prepare the company's budgets, long range plans, and outlooks to ensure financial goals are met.
- Provide variance analysis, oversight of the financial condition of the company to investor relations, and other senior management.
- Provide business managers and executives with accurate, timely, relevant, and insightful financial and operational information and results.

Career Background

Modern accounting was first popularized in the 1400's by Italian mathematician Luca Pacioli, when he popularized the double-entry bookkeeping system[129]. Luca introduced the use of journals and ledgers in accounting systems as well as the necessity of debits to equal credits[130]. Luca also introduced the concept of "closing the books", having year-end closing entries, and using ledgers to denote assets, liabilities, equity, income and expenditure accounts. The widespread use of double-entry accounting systems has been credited with allowing a free market system to develop rapidly, and capital to be allocated to its most highly valued use. The following quote denotes the contribution of Luca Pacioli:

> *"Capitalism designates an economic system significantly characterized by the predominance of "capital". Capitalism and double entry bookkeeping are absolutely indissociable; their relationship to each other is that of form to content." – Werner Sombart*

Accounting has been an integral part of the petroleum industry since its infancy. From how to finance exploration wells, to how to distribute revenues associated with oil & gas production back to their rightful owners, to payroll processes, accountants perform key processes in the on-going business for the oil & gas industry. Many captains of industry in oil & gas have had a background or strong ability for accounting.

Starting with John D. Rockefeller, when right after high school, he attended a 10 week business course and studied bookkeeping. Rockefeller used

---

[129] http://accountants-day.info/index.php/international-accounting-day-previous/77-luca-pacioli
[130] Ibid

accounting tools and techniques to develop the then largest oil & gas company in the world, Standard Oil:

"I charted my course by figures, nothing but figures." – John D. Rockefeller

Why you should be in Accounting or Finance

There are several benefits of having a career in accounting:

- If you have a strong interest in business, accounting or finance, having deep knowledge of these subjects can provide a solid foundation to learn more about a business through *financial transactions* that encompass the day-to-day operations of a company.
- Accounting is a highly portable profession and almost every industry or company requires a group of accounting employees.
- If you enjoy working with details, systems, numbers, and regulations.
- Accounting is the language of business and facilitates understanding the financial health of a company.
- Can lead to highly satisfying and high impact roles, such as Controller, Treasurer, Tax Officer, Chief Financial Officer, Vice President of Finance, and many other high level positions.

Why you should not be in Accounting or Finance

Following are a few cons of being in accounting:

- You like to think about the *big picture* and do not like worry about details. A good accountant needs to be very detail-oriented in order to succeed.
- Although not all assignments or positions within a finance function in a company entail a *repetitive* schedule every month, the majority of positions are impacted by *financial closing schedules* that are governed by internal and external requirements. If you like to have a completely flexible schedule, finance might not be your cup of tea.
- For most oil & gas companies, accounting & finance functions are commonly considered *cost centers*, which therefore limits the ability to have a fast career growth.
- If you would like to have a *direct profit and loss* responsibility, the finance function is usually relegated to a cost-cutting mode vs. more growth mode in most companies.

Career Development in Accounting & Finance

Most oil & gas companies have comprehensive programs for finance function employees. Typically, early career employees would start in transactional accounting roles that build a basic foundation not only in accounting, but also help early career employees acclimate to new systems, processes, organizational hierarchy, business processes, and overall flow. As an employee develops expertise, that employee may be then promoted or rotated to a financial analysis and reporting or FAR role. One of the biggest transitions for a university hire is the fact that positions do not have a specific guideline but may be more fluid, particularly in smaller companies where a financial analyst might be required to accomplish several responsibilities at the same time.

Educational & Skills Requirements

Depending on the role, the typical minimum degree required is an accounting, finance or economics degree. Although there are still non-degreed positions in the oil & gas industry, the number of those positions available is expected to be less and less over the next coming years. Most entry level positions require a bachelor's degree in accounting from an accredited university and if you have a degree in finance, many large companies require a *minimum* number of credit hours in accounting courses. In addition, large companies run *very structured* recruiting programs for Accounting & Finance where they typically hire from certain universities across the United States and they do the majority of hiring around the Fall semester. The reason for this is that budgets for companies are usually approved late in the year for next year's budget as well as many retirements occurring at the beginning of the following year.

Back Office vs. Front Office

The terms *back office* and *front office* are commonly used in financial services, trading operations, as well as the oil & gas industry. Front office consists of marketers, traders, business development and other personnel engaging in sale or trading activity while back office consists of support services, such as IT, accounting, HR and others. In the context of the finance function, the majority of the positions in Finance are catalogued as "back office" while there might be some finance support functions, which directly interact with front office employees such as traders or marketers that are labeled as "front office". For the purposes of this book, when referring to back office in the context of a finance employee this will refer to *transactional accounting* roles. In contrast, finance employees that have higher interaction with front level employees, such as Financial Analysis & Reporting, are considered *front office*.

Exposure

The finance function provides plenty of exposure to senior management in an oil & gas company. In an oil & gas company there are several skills that Finance that carry over from one function to the other:

- Attention to detail, with trying to tie out or match every number that is presented to another party.
- Excellent skills in Microsoft office applications, particularly Excel and PowerPoint.
- Inquisitiveness about how business transactions impact the financial statements and reporting.

There are several groups or functions that finance can provide an entry point to:

- Supply & trading of hydrocarbon commodities, with finance function employees being a career pathway into supply or trading.
- Petroleum marketing
- Business Development
- Market Analysis
- Investor Relations
- Chief Economics Roles

Possible long-term positions

Finance function based employees are considered for long-term positions such as:

- Chief Financial Officer
- Chief Accounting Officer or Comptroller
- Chief Tax Officer
- Treasurer
- Chief of Internal Audit
- Chief Executive Officer

Finance prepares people to have an impact on different organizations. Here are some examples of Finance-background employees who achieve Executive Management positions not finance-based:

- Jim Mulva, Former Chairman & CEO of ConocoPhillips, graduated with a degree in Finance and started in Finance.
- John Watson, Chairman & CEO of Chevron Corp, graduated with a degree in Economics and started in Finance at Chevron in 1980[131].
- Pierre R. Breber, EVP, Downstream & Chemicals of Chevron Corp, although an engineer by degree, he joined the Finance organization of Chevron in 1989[132].
- John E. Lowe, non-executive Chairman Apache Corp, prior EVP at ConocoPhillips, started in 1980 at Phillips Petroleum in the Finance organization.
- Gary R. Heminger, Chairman & CEO of Marathon Petroleum, earned a bachelors' degree in accounting and spent a significant percent of his time in the finance organization at Marathon[133].

Certified Public Accountant (CPA)

A Certified Public Accountant or CPA is a person who has passed the CPA exam, meets prior accounting experience, possesses an accredited accounting degree with enough credit hours and is licensed in one of the 50 U.S. states.

Unlike Public Accounting where a CPA is highly recommended and is actively sponsored by each accounting firm, in the oil & gas industry a CPA is only required in certain specific assignments. Here are some of the assignments or positions that often require or where a having a CPA would provide a *competitive advantage*:

- Income Tax compliance
- External Reporting under the Comptrollers' function
- Treasury management
- SEC compliance

---

[131] https://www.chevron.com/about/leadership/john-watson
[132] https://www.chevron.com/about/leadership/pierre-breber
[133] http://www.marathonpetroleum.com/About_MPC/Corporate_Profile/Corporate_Officers/Gary_R_Heminger/

Certified Financial Analyst (CFA)

The certified financial analyst designation is a very difficult designation to achieve in the world of Finance. The CFA is administered by the CFA Institute and consists of passing three exams. Level I exam is offered twice a year (June & December) while Level II and Level III exams are only offered in June. Getting the CFA takes a minimum of three years and requires extensive studying, with experts recommending studying more than 1,300 hours in total[134]. Level I requires approximately more than 350 hours of study time, Level II 500 hours, and Level III another 450 hours[135].

To become a CFA candidate, each candidate enrolls in the CFA program and in order to do that a candidate must have either have a bachelor's degree, have four years of professional experience or a combination of work experience and university experience totaling 4 years. The exam is conducted in English[136]. For more information visit www.cfainstitute.org.

Here are some of the careers or position or title that CFA members hold before or after completing the CFA:

- Relationship Manager
- Corporate Financial Analyst
- Financial Adviser
- Portfolio Manager
- C-level Executive
- Research Analyst

The CFA is highly geared and supported by large financial institutions, particularly equity research firms and others that may have exposure to oil & gas stocks. Other functions within the oil & gas industry that highly value CFA members are usually in the risk management, trading, business development, treasury functions, and many other departments of upstream, midstream & downstream companies.

---

[134] "Direct Path to the CFA Charter, by Rachel Bryant, CFA, Kindle edition, page 85
[135] Ibid
[136] https://www.cfainstitute.org/programs/cfaprogram/Documents/cfa_charter_factsheet.pdf

Examples of Careers in Finance

Transactional Accounting

As mentioned at the beginning of the chapter, transactional accounting roles provide a solid foundation of understanding business transactions as well as systems. Transactional accounting roles can work in a variety of locations, but often are concentrated in shared services centers, which could be local or global.

Some functions performed by Transactional Accounting functions:

- Production, Revenue & Royalty Accounting, which accounts for oil & gas production volumes, revenues and royalty and other regulatory reporting, associated with the upstream side of the business.
- Joint Venture Accounting, which account for joint operating agreements, billable costs to joint operating agreements, and billing other joint venture partners for expense and capital costs incurred in joint venture properties.
- Marketing & Trading Accounting, which account for the purchases and sales of various hydrocarbon products in the marketing or trading businesses.
- Accounts Receivable & Payable, which accounts receivable are reconciled, accounts payable are processed correctly and handle customers or vendors inquiries and concerns.

Production, Revenue & Royalty Accounting

Production, Revenue & Royalty Accounting is a predominantly U.S. Upstream function since *private* mineral rights ownership only exists in the U.S. There are more than one million oil & gas wells in the United States, so there is a significant amount of work that has to be performed by this function to correctly pay royalty and working interest owners around the country.

This function is usually responsible for:

- Recording and accounting for hydrocarbon production volumes for all of the company's operated and non-operated wells across many regions.
- Recording and accounting for hydrocarbon revenues, royalties and payments to the state, and other entities.

- Filing regulatory reports with agencies.
- Setting up and maintaining well information in the company's systems and keeping up with field changes, such as equipment changes, wells that are plugged and abandoned, or well recompletions.
- Dealing with ownership setup questions and troubleshooting any issues that may arise from royalty owners or other stakeholders.

Educational & Skills Requirements

As all finance and accounting positions most companies require, *at a minimum*, a bachelor's degree or equivalent in accounting or finance. Most exempt positions, or those that only require a high school diploma, have been dwindling in the recent years.

Besides a bachelors' degree, some background and knowledge of oil & gas production, leases, joint venture and other concepts is advantageous. In particular, analysts working in production accounting should become familiar with oil & gas production terminology. The COPAS accounting guidelines and books such as *"Oil & Gas Production in Nontechnical Language"* by Martin Raymond and Dr. William Leffler are good resources for somebody starting new into this field.

Day-to-Day Activities

The day to day activities vary from company to company, but can generally said to encompass activities tied to a "close schedule" for the first work days where all the accounting for oil & gas is completed. Please note that in the U.S., oil is accounted in the month following production month while natural gas is accounted with a one month lag. So for example, in March 2018 300 barrels of oil and 6,000 MCF or thousand cubic feet of natural gas were produced, the revenues for oil would be accounted for in April 2018 (Recorded in March 2018 for financial reporting purposes) while the gas will be accounted in May 2018.

Production Accounting:

- Setup well, equipment and other information as reported from field operations into the accounting system, prior to the well start producing so that volumes can be calculated correctly, reported to government agencies, and recorded on the company's books.

- Depending on the company, some production accountants receive *run tickets* or statements from oil transporters and these volumes then become the basis for sales for a well or a group of wells.
- Allocation of volumes from sales meters to producing wells, which allocate both sales and production for both oil and gas.
- File reports with regulatory agencies at the local, state, and federal level.
- Assist with royalty or working interest owners request as it regards to field setups.

Revenue Accounting:

- Setup contracts, certain allocation of deductions, and price formulas.
- Setup master data associated with oil and gas marketing contracts.
- Maintain *gas imbalances* for working interest owners taking gas in kind.
- Perform reconciliations to ensure revenues and royalty payments are accounted for correctly.
- Participate in *financial close* activities.

Systems

There are several systems that a production, revenue or royalty accountant would interact with:

- Production volumetric systems, both internal and external.
- Allocation systems.
- Enterprise Resource Planning Systems.
- Regulatory reporting systems.

The most popular Production Revenue Systems are the following:

- The PRA module of SAP, which the majority of large companies use since it integrates with SAP ERP software that houses other financial and operational processes.
- P2 Energy Solutions Enterprise Upstream, which is targeted to medium to large companies.
- P2 Energy Solutions Bolo, which includes also budgeting, forecasting, joint billing, as well production, revenue and royalty accounting.

In addition to those, analysts will have to interact with division of interest, who administer all the ownership records for the different oil & gas leases that the company owns.

Typical Job Titles
There are several job titles associated with this department:

- Production Analyst
- Revenue Analyst
- Royalty Analyst
- Severance Tax Analyst
- Owners Relation Analyst

Career Advancement
Since PRA is a *highly specialized* field, more likely promotions and career development will happen within that group. Career development philosophy varies significantly company to company; career advancement to other groups could be a challenge, especially when an employee is considered a subject matter expert in an area or highly key in a particular process. Careers in the field tend to be highly compensated and most companies have good career plans with employees choosing to specialize in this field. Possible positions outside of PRA include:

- Business unit financial analyst
- Royalty compliance
- Depending on experience, development program if employee is early career
- Joint Venture Audit

Portability
The opportunities in the PRA space are the *least* portable of all careers in Finance, since they are only available in companies with U.S. upstream operations. Since PRA is a highly specialized field that requires several years of experience to achieve true mastery as well as having complexity from the different state, local and company requirements that increase the challenges in this space. Employees with *deep expertise* and many years of experience are in high demand, particularly when oil & gas production increases as more wells being drilled require more analysts to account for all the hydrocarbons being produced. Individuals interested in this career path should always be

aware that for major companies there is a constant threat of *offshoring* to lower cost *service centers* in areas such as South America, Philippines or India.

Recommended Reading

There are several recommended readings, several of which are from the Council of Petroleum Accountant or COPA:

- Accounting Guideline 15 or AG-15 Gas Accounting Manual: This is a must have document for anyone involved in natural gas accounting. The purpose of the Gas Accounting Manual is to provide a vital reference for any accountant and others responsible for proper recording and reporting of natural gas from the wellhead to the point of sale and all related areas. Various areas addressed by AG-15 include production, measurement, processing, transportation, sales, settlements, regulatory compliance, as well as a history of regulatory compliance.
- AG-6 or Oil Accounting Manual: This publication provides generally accepted practices for oil allocations and related revenue accounting. It is intended to be of assistance to accountants and oil producers in establishing or revising oil accounting procedures and in training new personnel.

For more information visit COPAS' website at www.copas.org.

Marketing & Trading Accounting

Marketing accounting is primarily found in companies with downstream operations while trading accounting, depending on the commodity, can be found on upstream, midstream, and downstream operating companies.

Petroleum marketing operations around the world are highly complex, involve daily transactions, and the typical value of transactions is lower than what is seen in the trading space. Moreover, payment terms for the so-called *clean products* are very short, (normally less than 7 days) therefore invoice accuracy is critical. Another area that pressures marketing systems to be highly accurate and automated is the fact that the marketing of petroleum products is highly fragmented and competitive, no large company has more than 10% of the supply in a given area or region.

Marketing accounting is generally responsible for:

- Setting up customers and vendors in the company systems.

- Bill and invoice customers based on volumetric data confirmed with marketers and *petroleum storage & delivery terminals*.
- Apply cash payments to individual accounts, product groups and invoices.
- Reconcile transactions to ensure sales & accounts receivable are accounted for correctly and that there are no missing transactions.
- Resolve any logistical issues with customers and respond to billing and invoicing inquiries.

In the trading world, accounting is also highly complex and involves daily interaction with third party customers as well as internal customers like traders, schedulers, risk and other groups.

Trading accounting is usually responsible for:

- Setting up trading customers and vendors in company systems.
- Ensure that scheduled and delivered volumes and movements via modes of transportation such as barge, pipeline, railcar and others are accounted for both from a volumetric to a dollar perspective.
- Ensure volumes are recorded correctly in each step of the trading & scheduling process.
- Bill and invoice customers for commodities transactions, such as buys, sells and other as well as account for *secondary costs*, such as barge inspections, demurrage fees, transportation and other fees.
- Reconcile volumes and value within different operational and accounting systems, for sales, purchases and inventory.
- Troubleshoot any issues with the *deal capture* system and any questions traders or schedulers may have.
- Record, monitor and report on derivatives instruments used in commodities trading.

Educational & Skills Requirements
Similar to other transactional accounting roles, positions at either of these areas requires a bachelor's degree in accounting or finance.

Day-to-Day Activities
Day-to-day activities can vary among the different positions within this type of accounting, but can include:

- Billing customers for trading or marketing deals, which range from deals worth of millions in dollars, as in the case of say international

gasoline tankers, to less than 10,000 dollars for small gasoline station operators being billed for fuel received.
- Reconciling inventory movements throughout a company's assets, for product such as transportation fuels, crude oil, or NGLs.
- Setting up customers and vendors master data in various systems so that customers' payments can be received, vendors can be paid, and overall financial transactions recorded appropriately.
- Account for transportation fuels exchanges, a practice which is highly common in the oil & gas industry. Exchanges are used so that companies can fulfill contractual demands for a particular product without having to *physically* move that product. Let's say that Chevron has a need to supply gasoline in the Northeast but does not have supply there, but Shell does, Chevron and Shell could setup an exchange whereby Chevron delivers gasoline in Mississippi while Shell delivers gasoline in the Northeast so that both parties do not have to move *their own* products and avoid incurring transportation costs.

Systems

Marketing accounting deals with several systems such as:

- Customer Relationship Management systems, such as SAP CRM or Oracle.
- Terminal interface software.
- Enterprise Resource Planning systems such as SAP ERP or Oracle
- Customer billing systems, whether company proprietary or third party software companies.
- Volume tracking systems for trucking and pipelines.
- Pricing systems.

Trading accounting deals with multiple systems as well, including:

- Deal capture systems such as Solarc Right Angle, SAP Commodities Trading, Trilogy, and others. These systems are critical for trading operations since are akin to a "cash register" for a company.
- Logistics and scheduling systems that provide movement capture transactional data.
- Pipeline flow systems.
- Inventory reporting systems.

Typical Job Titles
Several titles include:

- Marketing Accountant
- Trading Accountant
- Trading Analyst
- Marketing Financial Services Supervisor
- Customer Master Data Setup
- Derivative Analyst
- Customer Service Analyst

Career Advancement & Portability
Positions within this function are relatively portable due to several reasons:

- Positions are often customer facing and interact with systems and processes not just used in the downstream business, but in any type of business that does trading or interacts with end-customers, be it small as in the petroleum marketing space, or larger companies as in the commodities trading space.
- Many of the skills learned through the trading business could be applied to other businesses outside the oil & gas industry that do commodities trading (agribusiness, manufacturers, banks, etc…).

Recommended Reading
The Association for Convenience & Petroleum Refining or NACS provides excellent reference documents, statistics, and news about the petroleum marketing world at the retail level. Their website is www.nacsonline.com.

The Petroleum Marketers Association of America (PMAA) is a federation of associations across the United States that represents more than 8,000 independent marketers around the US. Companies such as BP, Phillips 66, Chevron, Citgo, ExxonMobil, and many more are members of this association. Their website, www.PMAA.org, contains relevant industry statistics, studies, news and other insights available to the public about this great business.

To learn more about the trading business, we recommend a book titled "The Domino Effect" by Rusty Braziel. Rusty runs a widely-known consulting firm out of Houston called RBN Energy which also publishes a daily blog post around current topics around hydrocarbon markets.

Other Career Paths in Finance

Besides transactional accounting roles, there are many other careers in this field. Among some of the popular ones include:

- Financial Analysis & Reporting
- Internal Audit
- Tax

Financial Analysis & Reporting and Long Range Planning

Financial Analysis & Reporting, also called F&PA or FP&A, or business analysts is a set of roles within Finance & Accounting that deals more closely with analysis of key business drivers and provide financial reporting support.

Positions within this career path usually get involved in areas such as:

- Being a liaison between what gets reported in a company's financial systems and business participants. Key stakeholders in an oil & gas setting might include petroleum or chemical engineers, geosciences, management, marketing business, trading, or even other support staff groups such as HR, IT, HES or others.
- Coordinate a company's long range planning or budgeting exercise, from collecting key inputs, consolidate all the plans, help with projections or forecasts and present the company's budget to executive management.
- Provide *stewardship* or assistance in cost, revenue or inventory management and provide key financial insights to a company's decision makers.
- Support the company's financial regulatory filings, such as the *Management Discussion & Analysis* section of the Securities & Exchange Commission (SEC) form 10-K, Investor Presentations and other key external publications.

Key skills recommended for this career path:

- Good ability to communicate, particularly for complex financial transactions, into easy to follow business language without overuse of accounting *jargon* or intricate terminology.

- Ability to learn fast and juggle multiple requests at the same time. Because of the consistent deadlines and tight schedules around the *closing of the books*[137] , being able to *successfully* multitask is critical.

Long-term positions within this sub-function include VP of Finance, CFO, Manager of Financial Reporting, or General Manager of Finance.

Internal Audit

Internal provides a key role in making sure the company's business units and support groups comply with the company's internal control guidelines and maintain good control over financial reporting.

Internal Audit conducts a series of engagement for a company's business unit that provide reasonable assurance that internal controls are followed. Internal Audit's role has been elevated particularly since the passage of the Sarbanes-Oxley or SOX Act of 2002 after the corporate accounting scandals of the early 2000.

Some of the tasks that are performed by a company's internal audit team[138]:

- Verify the existence of assets and recommend proper safeguards for their protection;
- Evaluate the adequacy of the system of internal controls;
- Recommend improvements in controls;
- Assess compliance with policies and procedures and sound business practices;
- Assess compliance with state and federal laws and contractual obligations.
- Review operations or programs to ascertain whether results are consistent with established objectives and whether the operations or programs are being carried out as planned;
- Investigate reported occurrences of fraud, embezzlement, theft, waste, etc.

Tax

Despite common beliefs, Tax is one of the most diverse and all-encompassing functions within accounting & finance. Tax is a highly

---

[137] Closing the books are the series of activities in finance & accounting necessary to generate financial statements (income statement, balance sheet and cash flow) for a company or group of companies. In a large company, closing is accomplished through a series of interdependent tasks that are carried through many different people in possibly located in several countries.
[138] http://www.marquette.edu/riskunit/internalaudit/role.shtml

complex subject, particularly in the oil & gas industry, with far reaching implications into investment decisions, compliance, government regulations and lobbying. When most people think of tax they think of income tax, but as one of the most taxed industries in the world, there are a multitude of different types of taxes:

- Excise taxes: Excise taxes are levied all throughout the United States, Europe, and many other countries around the world. Excise and sales tax are particularly important for petroleum product sales, with these taxes representing anywhere from 20% of the value of the final product in *low tax* environments to more than 200% of the value[139]. Compliance and payments for these type of taxes is highly complex and carries significant penalties including revoking of the license to operate in certain jurisdictions[140].
- Sales and use tax: similar to individuals, corporations pay sales and use taxes around many jurisdictions not only in the United States but around the world. Sales taxes are levied on many of the equipment and inventory purchased by an oil & gas company and minimizing these tax liabilities is the goal of the sales and use tax compliance department. Since these types of taxes are unique to every county or even city in a country, the positions within this field are geared towards becoming a Subject Matter Expert and not simply a generalist.
- Property Tax: property taxes are similar in the United States to sales taxes in the sense that they vary across from each municipality to the other. Since oil & gas companies own or lease property in many areas across the country, the goal of this department is to comply with the property tax assessments that each location assesses on the various assets of an oil & gas company (for example refinery buildings, office buildings, oil & gas well, land for compressor stations, and many other types of real estate-like investments.
- State Income Tax: Most states in the U.S. have some sort of income tax that is levied on operations conducted within that state. The goal of this team in a tax department is to provide compliance and tax planning ability for state income tax filings across the country.

---

[139] Page 133, *"Oil & Gas Company Analysis: Petroleum Refining & Marketing"* 2017 edition
[140] For example, see Louisiana's Department of Revenue website: http://revenue.louisiana.gov/FAQ/QuestionsAndAnswers/3

- U.S. Federal Income Tax: This is the group that is most associated with the generic term Tax. The oil & gas industry, even after the recent U.S. income tax reform, is one of the most heavily taxed industries in the United States. This group deals with compliance, planning, and tax representation in all matters regarding this importance subject for companies.
- International Income Tax: Since the oil & gas business is a global business and the majority of companies have operations outside the US, they have income tax liabilities and compliance requirements for each country a company operates in. International income tax can be quite complex and requires the expertise of subject matter experts to be able to successfully comply with each country's extensive list of requirements.

Chapter VI – Engineering

"At its heart, engineering is about using science to find creative, practical solutions. It is a noble profession" – Queen Elizabeth II

What is Engineering?

The word engineer is derived from the Latin word *ingeniare* and *ingenium* which translates to invent, devise and cleverness. Engineering in a simple way is the application of science to solve problems. Engineers figure out *how* things work and find *practical* uses for scientific discoveries[141].

Engineering Disciplines

Engineering has many branches, but typically in the oil & gas industry we see the following disciplines:

- Civil
- Chemical
- Mechanical
- Electrical
- Petroleum

The oil & gas industry, as a highly complex, technical and critical industry, employs thousands of engineers all around the world in all sectors of the industry.

Civil Engineers

Civil engineers plan, design, construct, and maintain structures – such as buildings, roads, bridges, and dams – that meet human needs[142].

Chemical Engineers

Chemical engineers design equipment and processes for large-scale chemical manufacturing, plan and test methods of manufacturing products and treating byproducts, and supervise production. Chemical engineers also work in a variety of manufacturing industries other than chemical manufacturing, such as those producing energy, electronics, food, pet food, clothing, and paper. They also work in health care, biotechnology, and business services[143].

Mechanical Engineers

Mechanical engineers design, build and test machines, engines and, other mechanical devices[144].

---

[141] https://www.livescience.com/47499-what-is-engineering.html
[142] http://futuresinengineering.org/what.php?id=2
[143] http://futuresinengineering.org/what.php?id=2
[144] Ibid

Electrical Engineers

Electrical engineers design, develop, test, and supervise the manufacture of electrical and electronic equipment such as[145]:

- Broadcast and communications systems.
- Electric motors, machinery controls, and lighting and wiring in buildings, automobiles, aircraft, and radar and navigation systems.
- Power generating, controlling and transmission devices used by electric utilities.

Petroleum Engineers

Petroleum engineers are responsible for assessing the potential location, quantity and quality of hydrocarbon deposits or reservoirs as well for planning, managing, and optimizing production of hydrocarbons[146].

Career Development in Engineering

Engineers tend to be among the most widely deployed function within the oil & gas industry. Engineers are needed in all sectors of the industry and can work in variety of technical, supervisor and managerial roles. As an interesting fact, engineers often comprise the majority of executive positions in most energy companies.

Educational & Skills Requirements

Most if not all engineering positions today require a four or five year degree from an accredited university. Engineering degrees usually have a core set of classes centered on the natural sciences and math but largely the curriculum is focused on the engineer's discipline.

Some of the core subjects found in all engineering disciplines[147]:

- Introduction to Engineering
- Calculus I, II, & III
- Physics
- General Chemistry
- Engineering Ethics course
- Technical communications
- Statistics

---

[145] Ibid
[146] https://targetjobs.co.uk/careers-advice/job-descriptions/276301-petroleum-engineer-job-description
[147] http://www.chee.uh.edu/undergraduate/degree-plan

Depending on the engineering disciplines, some of the later courses in the degree will vary. The following table helps illustrate how the different disciplines share the same core courses and how the discipline courses differ:

	Civil	Chemical	Mechanical	Electrical	Petroleum
Core Courses		Intro to Engineering / Calculus / Ethics / Statistics / Physics / General Chemistry / Communications			
Discipline Courses	Physics / Engineering Math / Mechanics / Electricity and Optics / Civil Engineering Design / Materials Science / Fluid Dynamics / Thermodynamics / Geotechnical Engineering / Hydraulics	Chemical Lab / Materials Science / Organic Chemistry / Thermodynamics / Fluid Mechanics / Chemical E Design / Chemical Processes / Process/Model Control / BioChemistry / Chemical Reactions	Computer Aided Design / Fabrication / Advanced Physics / Materials Science / Heat Transfer / Fluid Mechanics / Thermodynamics / System Dynamics / Manufacturing / Solid Mechanics / Circuits & Electronics / Component Design / Mechanical Systems	Audio Technology / Circuit Design / Power Electronics / Control Systems / Power Systems / Photonics / Microelectronic Circuit Design / Waveguides and Antennas / Electromagnetic Waves & Devices / Signals & Systems	Physical Geology / Drilling Courses / Materials Science / Thermodynamics / Reservoir Engineering / Petrophysics / Reservoir Fluids / Reservoir Formation / Production systems / Reservoir Simulation / Well Testing

Technical Ladder vs. Management Ladder

Engineers in the oil & gas industry usually start in a technical ladder and depending on aptitudes, potential and possibly secondary education; they may be transitioned into a management ladder. Due to the technical nature of engineering, both the technical and management ladders, for most companies, are highly compensated and highly rewarding. A large factor in deciding whether a technical or management ladder is more appropriate depends heavily in terms of personality. Here are some questions to ask regarding the two:

- Do you value working in *hands-on* projects where you can have a high degree of autonomy?
- Can you manage *others* instead of being *self-sufficient*?
- Do you enjoy building something new, implementing and executing directly, and enjoy being a subject matter expert?

- How would you rate your emotional intelligence? If you have low emotional intelligence, you might be better suited to a technical ladder.
- Do you enjoy meeting about topics? Management ladder typically spends a significant portion of their time in business meetings.
- Do you enjoy developing others?
- Are you considered a mentor?
- Are you willing to defend your direct report's achievements and can allow them a degree of autonomy with checkpoints along the way?

Engineers that successfully transition from the initial technical ladder to a management ladder usually fit the following profile:

- An advanced master's degree, preferably an MBA from an accredited university. A significant percentage of CEO's in the oil & gas industry have an engineering degree coupled with an MBA degree, which allows an engineer to quickly learn the business and financial principles that an executive needs.
- Get promoted to supervisory positions relatively quickly in their careers.
- Volunteers or gets moved and complete a *cross-functional* assignment, where they pick up skills outside of their core competencies and broaden their experiences and understand the overall business.
- A good cross functional assignment could be in the company's trading operations, business development or other areas that have a direct *profit & loss* impact, which allows an engineer to transition from a *cost center* function to a *profit center* function. Additional applicable developmental assignments are in the finance department, financial evaluation, economic analysis, long range planning, strategy or investor relations.

Those engineers that prefer the technical ladder usually fit the following profile:

- Like to work alone or in a small team.
- Like solving complex problems.
- Enjoy spending hours working on something.
- Have little to no desire to lead a team.

Operating Company Engineer vs. Contracting or Vendor Company Engineer

One of the key decisions a young engineer in college has to make is whether they will be working for an *operating company* or a *vendor company*. Similar to the other functions where an employee is considered a *profit center* or a *cost center*, this determination will have long-term implications in an engineer's career.

Exposure

Engineering is one of the highest exposure careers in the oil & gas industry, primarily when working in large projects. Engineers in a new exciting or growth area can have access to local management which will improve their career potential long-term.

Possible long-term positions

As mentioned in prior chapters, the oil & gas industry is a highly complex, technically driven industry so it is not odd that engineers often tend to be the highest percentage of executives in this industry. This is particularly applicable in the exploration & production companies, where the Chief Executive Officer as well as the Executive Vice Presidents tend to have an engineering background.

The following are recent executives who have an engineering degree and who have led major oil & companies:

- Greg Garland, Chairman & CEO of Phillips 66, received a bachelor degree in chemical engineering from Texas A&M University in 1980[148]. He started his career with Phillips Petroleum progressing through engineering and later management roles. He went onto lead the company's chemical JV company, ChevronPhillips Chemicals.
- Darren Woods, Chairman & CEO of ExxonMobil, received a bachelor degree in electrical engineering from Texas A&M. He joined Exxon in 1992 and progressed through a series of domestic and international assignments[149].
- John Christmann is the CEO and President of Apache Corporation, received his bachelor degree in petroleum engineering from the Colorado School of Mines as well having an MBA. He

---

[148] Source: Phillips66.com
[149] http://corporate.exxonmobil.com/en/company/about-us/management/darren-w-woods

joined Apache in 1997 after having a career with Vaster Resources/ARCO Oil and Gas[150].
- Mark Papa, founder and former Chairman and CEO of EOG Resources, received his bachelor degree in petroleum engineering from the University of Pittsburgh. He started his career at Belcom Petroleum and in 1999 he founded EOG Resources.

Sample positions

There are several long-term positions in Engineering, as well as the discussed executive positions:

- Functional engineering manager, whereby an engineer manages all the technical engineers in a project or department. Engineering managers would tend to focus on not just applying their technical skills, but the ability to lead, motivate and manage highly technical individuals, promote their career, and become a good stable foundation for the team. These positions are highly rewarded since they impact highly critical roles in a company (for example, the reservoir manager who manages or coaches several petroleum engineers who specialize in the field of reservoir management and hydrocarbon production optimization.
- Project managers, with engineers comprising a very large percentage of project managers in the oil & gas industry. Engineering project managers can come from all disciplines and often take project management courses or even the coveted *Project Management Professional certification* or PMP, which is covered in the next few pages.
- Lead in another discipline, such as trading, business development, or even accounting. Many individuals with a background in engineering had a cross functional assignment in another discipline, such as trading operations or logistics, and did not come back to a core engineering area, with having a lead position in those fields. The experience and knowledge gained in initial engineering positions served these individuals well; they did not have to return to their core discipline.

Professional Engineering Certificate

There are a variety of degrees within engineering that require having a professional engineering or P.E. certificate. Professional Engineer

---

[150] http://www.apachecorp.com/About_Apache/Management/John_J_Christmann_IV.aspx

designations are handled at the state level in the United States and typically by national agencies in foreign countries.

In order to use a P.E. designation, engineers must complete several steps to ensure their competency[151]:

- Earn a four-year degree in engineering from an accredited engineering program.
- Pass the Fundamentals of Engineering (FE) exam.
- Complete four years of progressive engineering experience under a PE.
- Pass the Principles and Practice of Engineering (PE) exam.

There are several benefits of becoming a licensed professional engineer[152]:

- Only a licensed engineer may prepare, sign and seal, and submit engineering plans and drawings to a public authority for approval, or seal engineering work for public and private clients.
- Licensure for engineers in government has become increasingly significant. In many federal, state, and municipal agencies, certain governmental engineering positions, particularly those considered higher level and responsible positions must be filled by licensed professional engineers.
- Many states require that individuals teaching engineering must also be licensed.

Chemical Engineering

Chemical engineers translate processes developed in the lab into practical applications for the commercial production of products and then work to maintain and improve those processes. They rely on the main foundations of engineering: math, physics, and chemistry (though biology is playing an increasing role). The main role of chemical engineers is to design and troubleshoot processes for the production of chemicals, fuels, foods, pharmaceuticals, and biologicals, just to name a few. They are most often employed by large-scale manufacturing plants to maximize productivity and product quality while minimizing costs.

---

[151] https://www.nspe.org/resources/licensure/what-pe
[152] Ibid

The aerospace, automotive, biomedical, electronic, environmental, medical, and military industries seek the skills of chemical engineers in order to help develop and improve their technical products, such as:

- Ultrastrong fibers, fabrics, and adhesives for vehicles.
- Biocompatible materials for implants and prosthetics.
- Films for optoelectronic devices.

Chemical engineers work in almost every industry and affect the production of almost every article manufactured on an industrial scale. Some typical tasks include[153]:

- Ensuring compliance with health, safety, and environmental regulations.
- Conducting research into improved manufacturing processes.
- Designing and planning equipment layout.
- Incorporating safety procedures for working with dangerous chemicals.
- Monitoring and optimizing the performance of production processes.
- Estimating production costs.

Educational & Skills Requirements

Chemical engineers must have a bachelor's degree in chemical engineering or a related field. Programs in chemical engineering usually take four years to complete and include classroom, laboratory, and field studies. High school students interested in studying chemical engineering will benefit from taking science courses, such as chemistry, physics, and biology. They also should take math courses, including algebra, trigonometry, and calculus.

At some universities, students can opt to enroll in 5-year engineering programs that lead to both a bachelor's degree and a master's degree. A graduate degree, which may include a degree up to the Ph.D. level, allows an engineer to work in research and development or as a postsecondary teacher.

Below are some of the skills required for chemical engineers to succeed in the workplace[154]:

---

[153] https://www.acs.org/content/acs/en/careers/college-to-career/chemistry-careers/chemical-engineering.html

- Analytical skills. Chemical engineers must troubleshoot designs that do not work as planned. They must ask the right questions and then find answers that work.
- Creativity. Chemical engineers must explore new ways of applying engineering principles. They work to invent new materials, advanced manufacturing techniques, and new applications in chemical engineering.
- Ingenuity. Chemical engineers learn the broad concepts of chemical engineering, but their work requires them to apply those concepts to specific production problems.
- Interpersonal skills. Because their role is to put scientific principles into practice in manufacturing industries, chemical engineers must develop good working relationships with other workers involved in production processes.
- Math skills. Chemical engineers use the principles of advanced math topics such as calculus for analysis, design, and troubleshooting in their work.
- Problem-solving skills. In designing equipment and processes for manufacturing, these engineers must be able to anticipate and identify problems, including such issues as workers' safety and problems related to manufacturing and environmental protection.

Systems

Some of the typical systems used by chemical engineers in the oil & gas industry are:

- Aspen HYSIS, is a leading process simulation software, used in a variety of businesses, from multiflow simulation in pipelines to refineries planning on how to optimize a refinery. Some of the tasks this tool can perform are[155]:
    - Plant design and turnaround optimization
    - Plant-wide margin analyses to boost profitability
    - Safety and operability studies
    - Equipment modeling and economics analyses
    - Energy efficiency optimization

Typical Job Titles

- Process Engineer

---

[154] https://www.bls.gov/ooh/architecture-and-engineering/chemical-engineers.htm#tab-4
[155] http://home.aspentech.com/products/engineering/aspen-hysys

- Optimization Engineer
- Unit Engineer
- Process Optimization Engineer

Career Advancement

Chemical engineers are some of the flexible careers within the oil & gas industry, being able to work in all three sectors plus the service sectors as well. Chemical engineers usually start by being a process engineer for a small subprocess in a refinery or even in the E&P world by being in charge of a small subject of processes or production area. Usually in a downstream environment, a chemical engineer could start as process engineer, then unit engineer, then process manager and even become refinery manager. Other career areas include specializing in natural gas processing or gathering, modeling pipeline fluid dynamics for a midstream asset, or even becoming project managers. In the upstream side, chemical engineers frequently concentrate in the production and treating facility side of the business, managing projects related to tanks, pumps, pipelines, and separators[156]. Chemical engineers can have quite rewarding careers in this industry.

Compensation

Chemical engineers are the most highly compensated engineering disciplines in the oil & gas industry, just behind petroleum engineers, which are discussed further in the chapter. The top 10 chemical engineers typically earn more than $140,000 per year according to a recent survey[157].

Portability

Chemical engineers have some of the most portable careers within the oil & gas industry as well as outside the industry. Chemical engineers can easily work in the upstream, midstream, or downstream sectors. Moreover, for those people interested in portability to other industries, a career in chemical engineering provides a very good balance between compensation and portability.

---

[156] https://www.aiche.org/chenected/2010/12/where-do-chemical-engineers-fit-upstream-oil-and-gas-industry
[157] https://bizfluent.com/info-8673266-benefits-being-chemical-engineer.html

Recommended Reading

There are several recommended reading resources for readers interested in a career in this field:

- The Bureau of Labor Statistics or BLS publishes an occupational handbook which details many career fields, including chemical engineering.
- For those interested in working in the downstream side of the business, we highly recommend getting a copy of *Oil & Gas Company Analysis: Petroleum Refining & Marketing* by Alfonso Colombano, which provides a comprehensive overview of the downstream business from a financial perspective.
- To better understand how a refinery works and their different processes, we recommend readers get a copy of *Petroleum Refining in Nontechnical Language* by William Leffler (published by PennWell).
- For a free resource on refining processes, readers can go visit OSHA's petroleum refining processes at www.osha.gov.

Electrical Engineering

An electrical engineer is someone who *designs* and *develops* new electrical systems, solves problems, and tests equipment. They study and apply the physics and mathematics of electricity, electromagnetism, and electronics to both large and small scale systems to process information and transmit energy. Electrical engineers are usually concerned with large-scale electrical systems such as motor control and power transmission, as well as utilizing electricity to transmit energy[158].

Electrical engineers design, develop, test, and supervise the manufacturing of electrical equipment, such as electric motors, radar and navigation systems, communications systems, or power generation equipment. Electrical engineers also design the electrical systems of automobiles and aircraft.

Educational & Skills Requirements

Electrical engineers usually earn their 4-year college degree from a variety of accredited universities in the US and around the world.

---

[158] https://www.engineering.unsw.edu.au/electrical-engineering/what-we-do/what-do-electrical-engineers-do

Below is a good starting list for some of the skills required[159]:

- Concentration. Electrical engineers design and develop complex electrical systems and electronic components and products. They must keep track of multiple design elements and technical characteristics when performing these tasks.
- Initiative. Electrical engineers must apply their knowledge to new tasks in every project they undertake. In addition, they must engage in continuing education to keep up with changes in technology.
- Interpersonal skills. Electrical engineers must work with others during the manufacturing process to ensure that their plans are implemented correctly. This collaboration includes monitoring technicians and devising remedies to problems as they arise.
- Math skills. Electrical engineers must use the principles of calculus and other advanced math in order to analyze, design, and troubleshoot equipment.
- Speaking skills. Electrical engineers work closely with other engineers and technicians. They must be able to explain their designs and reasoning clearly and to relay instructions during product development and production. They also may need to explain complex issues to customers who have little or no technical expertise.
- Writing skills. Electrical and electronics engineers develop technical publications related to equipment they develop, including maintenance manuals, operation manuals, parts lists, product proposals, and design methods documents.

Typical Job Titles

There are several job titles associated with being an electrical engineer in an oil & gas company, including:

- Instrumentation & Electrical Engineer
- I&E Supervisor
- I&E Specialist
- I&E SME

---

[159] https://www.bls.gov/ooh/architecture-and-engineering/electrical-and-electronics-engineers.htm#tab-4

Career Advancement

Electrical engineers, similar to mechanical engineers, have a wide flexibility in not only working in the different sectors within the oil & gas industry, but also in working for other industries. Typically, electrical engineers would start working with a more senior engineer and then progress throughout their career:

- I&E Engineer Junior
- I&E Engineer Senior
- Those following the management path career ladder in a corporation would frequently venture out to become a supervisor within the first 5-7 years of their career
- I&E Supervisor
- I&E Superintendent

There are also many electrical engineers that become skilled project managers and do not necessarily have to become a *Subject Matter Expert* in their own field per se.

Compensation

The average salary for electrical engineers across all disciplines is around $100,000 as per a recent survey from the Bureau of Labor Statistics[160]. Within the oil & gas industry, electrical engineers working in offshore or international projects would tend to command the highest rate of compensation, particularly those that can benefit from expat assignments. Within the downstream space, compensation tends to be slightly lower than upstream, particularly because of the *base operations* sense for most refineries in comparison to the *project driven* investment phases in upstream. For engineers in the midstream space, compensation would typically be the lowest due to the *lower complexity* of assets, particularly in the pipeline business from an Instrumentation and Electrical function point of view.

Portability

Along with mechanical engineers, electrical engineers can highly benefit from extended career portability, particularly in comparison to petroleum and to a lesser extent chemical engineers. An experienced, fast learner and dedicated electrical engineer can switch industries since many other

---

[160] https://www.bls.gov/ooh/architecture-and-engineering/electrical-and-electronics-engineers.htm#tab-5

industries require I&E expertise, not those just in the oil & gas industry, but also in manufacturing, construction, aviation, power, to name a few.

Mechanical Engineering

Mechanical engineering provides some of the most portable concentrations in engineering. Mechanical engineers can work in almost any type of industry that has rotating equipment. Mechanical engineering can be thought of as the *"successful* and *commercial* application of applied physics". Mechanical engineers, by using physics, mathematics, mechanics and other concepts, apply these scientific principles in a business and commercial space.

Mechanical engineers generally do the following[161]:

- Analyze problems to see how mechanical and thermal devices might help solve a particular problem.
- Design or redesign mechanical and thermal devices or subsystems, using analysis and computer-aided design.
- Develop and test prototypes of devices they design.
- Analyze the test results and change the design or system as needed.
- Oversee the manufacturing process for the device.

Mechanical engineers design and oversee the manufacture of many products ranging from medical devices to new batteries.

Mechanical engineers design power-producing machines, such as electric generators, internal combustion engines, and steam and gas turbines, as well as power-using machines, such as refrigeration and air-conditioning systems.

Mechanical engineers design other machines inside buildings, such as elevators and escalators. They also design material-handling systems, such as conveyor systems and automated transfer stations.

Like other engineers, mechanical engineers use computers extensively. Mechanical engineers are routinely responsible for the integration of sensors, controllers, and machinery. Computer technology helps mechanical engineers create and analyze designs, run simulations and test how a machine is likely to work, interact with connected systems, and generate specifications for parts.

---

[161] https://collegegrad.com/careers/mechanical-engineers

Educational & Skills Requirements

Mechanical engineers should have the following qualities[162]:

- Creativity. Mechanical engineers design and build complex pieces of equipment and machinery. A creative mind is essential for this kind of work.
- Listening skills. Mechanical engineers often work on projects with others, such as architects and computer scientists. They must listen to and analyze different approaches made by other experts to complete the task at hand.
- Math skills. Mechanical engineers use the principles of calculus, statistics, and other advanced subjects in math for analysis, design, and troubleshooting in their work.
- Mechanical skills. Mechanical skills allow engineers to apply basic engineering concepts and mechanical processes to the design of new devices and systems.
- Problem-solving skills. Mechanical engineers need good problem-solving skills to take scientific discoveries and use them to design and build useful products.

Typical Job Titles

Mechanical engineers can work in a variety of positions in the oil & gas industry, but some of the most typical job titles for early to mid-career mechanical engineers:

- Maintenance Engineer
- Operations Engineer
- Reliability Engineer
- Integrity Engineer
- Rotating Equipment Engineer

Career Advancement

Mechanical engineers can be found in all sectors of the oil & gas industry and they typically tend to work in operations & maintenance areas:

- Reliability Engineer, whereby an engineer works in analyzing maintenance and operations data and preparing preventative maintenance plans.
- Maintenance Supervisor.

---

[162] https://collegegrad.com/careers/mechanical-engineers

- Maintenance Superintendent.
- Operations Superintendent.
- Plant or District Manager.
- Refinery Manager.
- Vice-President Operations or Refining.

Compensation

According to a recent BLS survey, mechanical engineers median salary is around $100,000, but this is the median of all industries. Within the oil & gas industry, particularly the upstream space, median salaries are around $125,000-$150,000 per year plus benefits. Mechanical engineers employment tends to be more stable in process industries that need to have high reliability such as LNG plant, refineries, fractionators, and other "must-run" facilities that require an extensive staff of experienced mechanical engineers to address any rotating equipment or maintenance issue.

Portability

Mechanical engineers have one of the most portable and applicable careers in and outside the oil & gas industry. Mechanical engineers are needed in every asset of an oil & gas company, from drilling operations, to production, to pipeline operations, to refineries to marketing operations. Mechanical engineers often work in maintaining the operating reliability of a facility, minimize maintenance costs, and prolong the existing life of the asset as much as economically possible.

Petroleum Engineering

Petroleum engineering is one of the most highly rewarding fields in the oil & gas industry. Petroleum engineering can be thought of a mixture of chemical engineering plus geology.

Petroleum engineers work with teams of specialists to develop more effective, cost-efficient methods of producing oil & gas through the application of principles from chemistry, mathematics, engineering, and geology. There are three types of petroleum engineers who differ only by the stage of the drilling process that they are involved in. Reservoir engineers monitor the geological formation for the best strategic method of extraction. Drilling engineers generate computer simulated models of the drilling formation and equipment to ensure they use the best tools for the most effective method of extraction. Production engineers manage the interface between drilling and extraction by managing machinery and production costs[163].

When a new reservoir is located, petroleum engineers analyze it to determine whether it can be profitably exploited. If so, they create a drilling and extraction plan to pump out the oil or gas with as little cost as possible. Often, reservoirs have internal divisions, and the engineer must find the most efficient way to drill through those walls so the oil will flow freely to a single well. With shale oil, engineers calculate how to fracture the shale beds most efficiently to free the gas or oil and extract it[164].

From a location perspective, petroleum engineer has the most diversity by company. Some exploration & production companies have a central engineering department that handles petroleum engineering requirements versus at the local site, but most P.E. would have to visit the site's operations several times a year. In the United States and most countries abroad, the location of petroleum producing activities is not located in the most comfortable locations. Other companies have localized field offices where all personnel, including engineering, are located at.

Educational & Skills Requirements

Similar to other engineering disciplines, most if not all petroleum engineers need to have a 4 or 5 year degree from an accredited university. Most entry level positions in reservoir, production or drilling engineering prefer to have a specific degree in petroleum engineering, although certain engineering disciplines can transition into these roles, such as chemical or mechanical engineer with taking specific courses.

Systems

Several systems are used by petroleum engineers, including:

- Schlumberger's Reservoir Engineering software, which allows to run reservoir simulations for all types of crude oil.
- Schlumberger's drilling planning and operations is a software package covering several individual software, such as Drillbench, petrel drilling, and others[165].
- Petrel E&P Platform is a software that brings several disciplines together, such as geologists, geophysicists, and petroleum engineers

---

[163] https://study.com/articles/Petroleum_Engineer_Job_Description_and_Info_About_a_Career_in_Petroleum_Engineering.html
[164] http://work.chron.com/typical-duty-petroleum-engineers-10436.html
[165] https://www.software.slb.com/products/disciplines/drilling-software

as an overall environment that brings together exploration to production of hydrocarbons[166] to increase productivity. This software captures knowledge all the way from petroleum systems modeler to the reservoir engineer and beyond, covering all aspects of the subsurface.

- Pipesim, which is a multiphase flow simulation software that allows engineers to design production and gathering lines to optimize production from oil & gas wells.
- Oilfield Manager (OFM) is a well and reservoir analysis software that allows an engineer to monitor and survey performance of the wells, and forecast production with decline and type curve analysis. This software also allows viewing, relating and analyzing reservoir and production data with comprehensive tools, including interactive base maps with production trends, bubble plots, and diagnostic plots[167].
- Fracpro, this software can model any type of pressure stimulation job, particularly those used in the shale wells. This software can model how fractures can grow in any formation, giving a better understanding of how to place *proppant*[168], well productivity improvements, and fracture dimensions[169].
- IP, a subsurface solution that helps petroleum engineers determine the amount of hydrocarbons in the reservoir.

Typical Job Titles

Industry standard titles are as follows:

- Production Engineer
- Reservoir Engineer
- Drilling Engineer
- Drilling Superintendent
- Reservoir Manager

---

[166] https://www.software.slb.com/products/petrel
[167] https://www.software.slb.com/products/ofm
[168] A proppant is a material, commonly ceramic or sand-based, that is designed to keep a hydraulic fracture (i.e. shale wells) open during or following a fracturing treatment. The ability to determine which proppant to use is critical to increase productivity of wells. Without proppant, the shale revolution would not have occurred.
[169] http://www.carboceramics.com/Oil-gas/fracpro/FRACPRO-fracture-design-and-analysis-software

Career Advancement

The following are examples of types of petroleum engineers[170]:

- Completions engineers decide the best way to finish building wells so that oil or gas will flow up from underground. They oversee work to complete the building of wells, which might involve the use of tubing, hydraulic fracturing, or pressure-control techniques.
- Drilling engineers determine the best way to drill oil or gas wells, taking into account a number of factors, including cost. They also ensure that the drilling process is safe, efficient, and minimally disruptive to the environment.
- Production engineers take over wells after drilling is completed. They typically monitor wells' oil and gas production. If wells are not producing as much as expected, production engineers figure out ways to increase the amount being extracted.
- Reservoir engineers estimate how much oil or gas can be recovered from underground deposits, known as reservoirs. They study reservoirs' characteristics and determine which methods will get the most oil or gas out of them. They also monitor operations to ensure that the optimal levels of these resources are being recovered.

Compensation

Petroleum engineers the most highly compensated engineers in the oil & gas industry. Petroleum engineers, depending on experience can earn up to $140,000 per year, depending on the company.

Portability

Petroleum engineering is the *least portable* of all the engineering disciplines, even between the different sectors of the industry. Petroleum engineering is highly correlated with the level of activity in exploration and production, with times of high prices having an increased demand and compensation for engineers. On the other hand, as a trade-off to this high level of compensation in the upstream, petroleum engineers have the *least career* portability of all engineering disciplines. A mechanical engineer can work in pretty much any industry, the same for chemical, electrical and civil engineers while petroleum engineers can mostly only work in upstream oil & gas.

---

[170] https://collegegrad.com/careers/petroleum-engineers

Recommended Reading

We recommend the following materials to increase knowledge in the field of petroleum engineering:

- Guidelines for the Evaluation of Petroleum Reserves and Resources[171].
- Petroleum Resources Management System, published in 2007, provides a comprehensive guide to the basis that becomes how companies publish and report petroleum reserves around the world[172].

Another good reading material comes from the U.S. Securities and Exchange Commission, which regulates how specifically U.S. listed companies can report the different concepts of petroleum reserves, how proved reserves should be disclosed, and many other concepts. This guide is called "SEC Final Rule: Modernization of Oil and Gas Reporting"[173]. These rules were adopted in 2008 after oil prices experienced a significant increase throughout the summer and then fell down precipitously, calling to question the prior formula of using the last-day of the year to estimated reserves.

---

[171] http://www.spe.org/industry/docs/GuidelinesEvaluationReservesResources_2001.pdf
[172] http://www.spe.org/industry/docs/Petroleum_Resources_Management_System_2007.pdf
[173] https://www.sec.gov/rules/final/2008/33-8995.pdf

Chapter VII – Other Career Groups

"You can make a lot of mistakes and still recover if you run an efficient operation. Or you can be brilliant and still go out of business if you are too inefficient" – Sam Walton

Operations

> *"Hire people who are better than you are, then leave them to get on with it. Look for people who will aim for the remarkable, who will not settle for the routine."* – David Ogilvy

Operations personnel can work in a variety of fields in the oil & gas industry. As of the time of this writing, the *shortage* of skilled trades continues to *worsen* in the United States and other developed countries[174]. Due to this *skills* shortage, the compensation for positions within operations that require *technical skills* or on the job skills *not typically learned* in a university setting is rising at a *faster pace* than many other jobs. In many cases compensation for skilled trades is significantly *higher* than those with university degrees with limited marketability.

Described below are some of the career opportunities available within the very large space of operations:

- Operations & maintenance personnel in a downstream oil refinery perform a wide variety of operations and maintenance tasks, such as running the facility to minimize downtime, conduct safety procedures before executing any work on equipment (i.e. LOTO[175], JSA or Job Safety Analysis), and executing maintenance requests.
- Upstream operators assure the continuous operations of upstream wells, and perform activities such as metering of volumes, calibration, maintenance on the wells, and operational reporting in the field. Operators also assure that the custody transfer and measurement process goes as expected and minimizes loss of hydrocarbons through incorrect measurements or oversights (i.e. onshore crude oil run tickets). Lastly, operators provide key data analytics for a company to analyze the on-going performance and integrity of a company's oil & gas wells.
- Transportation operations personnel transport hydrocarbons or hydrocarbon feedstocks from one site to the other, whether they are crude oil haulers, fuels truckers, vessel crew or in any other function moving hydrocarbons along the value chain. Transportation personnel, particularly, in the trucking field in the

---

[174] http://www.theauditoronline.com/skilled-worker-shortage-most-anticipated-challenge-of-2018/
[175] LOTO stands for Lock-Out Tag-Out, and it is a basic safety procedure before doing any work on any equipment in a facility. For example, you would not want for an employee to disconnect a running steam pipe and be exposed to high temperature steam. The LOTO procedure is performed in any type of facility that has equipment, whether they are in the oil & gas industry or not.

United States, are quite critical in terms of assuring that crude oil gets from the many, many wells across the country to crude terminals and then transport this crude oil to refineries or to export terminals for overseas markets.

- In construction, operations personnel execute construction plans and perform a variety of tasks, from digging earth to laying pipelines, to pouring concrete to building foundations for a new fractionator, to building scaffolding for other workers, to welding. Many of the critical activities that are key in constructing a new asset are performed by operations personnel.

Educational & Skills Requirements

Educational requirements vary from company to company, but the majority of companies require *at a minimum* a high school degree and preferably an associate's degree, or a degree from a trade or technical school.

Those looking for employment in particular fields, such as marine or on-shore transportation are usually required to have licensing or certification such as:

- U.S. Department of Homeland Security transportation security card.
- Marine licensing
- Commercial Driver's License or CDL
- Construction specific licensing
- Safety certifications

There are several skills that are generally thought to help in working in operations:

- Ability to have good *eye*, *hand*, and *muscle* coordination.
- Being able to work outdoors without issues, particularly in the summer months in warmer environments.
- Ability to translate abstract concepts, such as a construction diagram, into actionable steps in the physical world (i.e. welding).
- Being able to lift a certain amount of weight without assistance and in general having a physical ability to be able to execute work effectively.
- Enjoy standing in risky areas (i.e. scaffolding) for prolonged periods of time without any impacts to health.

- Having good driving ability and good driving record, essential for those considering a position in transportation.
- As the industry increases the use of technology, being adapt at learning new systems is increasingly becoming more and more critical than ever before.

Typical Job Titles

There are many, many job titles associated with working in an operations in this industry, below are just a few of the most commonly used. Please note that many companies would have slightly different terminology, but these titles would approximately perform similar work:

- Multi-Skilled Operator or simply operator would be the personnel responsible for the safe, reliable and optimal production from upstream wells.
- Maintenance associate, specialist or analyst, personnel involved in planning and executing maintenance tasks, whether they are *preventative*, *corrective*, or even *predictive* maintenance.
- Transportation personnel, such as vessel captain, trucker or hauler, which assist in transporting hydrocarbons from one location to another.

Career Advancement & Stability

In general career advancement depends heavily on the type of company and the focus that a particular company provides to developing its internal pool of *non-university degreed* personnel. Many companies, even the larger ones, have realized how critical the skillsets and impacts these employees can have in a company and have started to develop career management groups for these groups of employees.

Depending on how exposed to a particular commodity a company might be, the *more* or *less* stable many positions within this field will be. As a general comment it can be said that:

- Positions that are geared towards *maintaining* or *growing* production, such as operators in a hot growing play like the Permian basin currently, are safer.
- Operator positions in basins that have fallen out of favor, such as currently Wyoming, San Juan basin in New Mexico are *less stable*.
- Assets, such as Caribbean refineries, central European refineries, or Japanese refineries, which face either high energy costs, labor costs

or are simply not placed in the most strategic position to sustain profitable operations, can be at risk for layoffs.

There are several looming technology threats that could impact career stability for these positions:

- Advancement of robotic process automation might reduce the available opportunities for people interested in this field. Even fields that have been in the past out of reach for automation purposes such as truck driving are now being considered for automation. In fact, many experts predict that within 10 years a full driverless truck might become a reality as recent experiments have demonstrated[176].
- To the extent a position requires less physical work, the higher the threat of offshoring to lower cost environments. For example, through the use of "service centers" located outside the U.S., many large companies do not need any longer operations assistance personnel that do manual computer based work such as entering crude oil run tickets in a data capture system, instead relying on the operators to scan these tickets and have an offshore center key in and validate data. Many companies are even simply automating the data capture and thus not even have to rely on manual data entry.

Compensation

Compensation within this field can vary wildly, not only from company to company, but within the different options and also within the different locations across the U.S. and around the world. In general, larger companies tend to compensate their technical operations personnel with more long-term benefits such as 401k matching, pension plan schemes, and better insurance benefits. The smaller companies tend to focus on providing higher cash or salary compensation, but tend to provide fewer benefits but also a higher degree of autonomy and less reliance on established processes and procedures than the larger companies.

Portability

The portability of skills gained through exposure in the vast array of operations careers depends widely on specific the skill is. For example, the skills and experience gained by a welder are more widely applicable than say

---

[176] https://hackernoon.com/starskyunmanned-de7af7e5a3

a wet gas measurement specialist, which is only transportable within upstream or midstream operations.

Geosciences

"The field of the Geologist's inquiry is the Globe itself, and it is his study to decipher the monuments of the mighty revolutions and convulsions it had suffered" – *William Buckland*

The field of geosciences lies at the very essence of finding and developing hydrocarbons. Without the necessary skills from the many careers within geosciences oil & gas deposits would not be discovered and thus produced from these reservoirs.

Educational & Skills Requirements

Most positions require at a minimum a bachelor's degree in one of the different disciplines within geosciences, such as geology or geophysics. With increased competition in the business, most companies require masters or sometimes even a Ph.D. degree.

The geosciences fields require many different skills, including:

- Good understanding of petroleum geology, how hydrocarbons are formed and discovered.
- Good math and physics skills, with this field being highly technical and more oriented towards becoming a *subject matter expert*.
- Ability to think in *abstract terms* and visualize the *subsurface*. While other disciplines such as chemical or mechanical engineering work with *visible* materials and products, geosciences, much like petroleum engineers, have to work with widely unknown and have to analyze data from core and imaging that is located *thousands* of feet underground.

Typical Job Titles

- Junior Geologist
- Exploration Geologist
- Senior Geophysics
- Exploration Manager/Supervisor
- VP of Exploration
- Geophysicist
- Geological Research

Career Advancement & Stability

Career progression is similar to other science-based fields, but with a higher percentage of SME's and *less* cross functional assignments within other positions.

Stability largely depends on the size of the company with more established companies having more stable but lower compensated geoscience positions.

One of the highest positions a geoscientist can achieve in an E&P company is to become the "Chief Geologist" or VP of exploration, which are two of the highest positions within the field. A chief geologist, working very closely with the petroleum reserves department, is responsible for evaluating and reviewing the quality of the company's resources, reserved and proved reserves, the absolute key asset in an upstream company. The reserves attestation is filed with Securities & Exchange Commission in the United States and other jurisdictions.

Compensation

Geosciences tend to be amongst the most highly compensated fields in the oil & gas industry. Geoscientists working in smaller companies would tend to have a higher percentage of their compensation tied to either company stock or some other type of rewards so that they can receive a lower salary in the short term. For example, if more hydrocarbons are discovered than expected, then that geoscientist would have more company stock that could be highly rewarding if that smaller company eventually goes public.

Portability

The entire field of geosciences, along with petroleum engineering, is one of the *least portable* fields within the oil & gas industry. Not only are geosciences personnel limited to the upstream oil & gas industry, even transferring those skills to other sectors, such as midstream or downstream within the industry remains a challenge. Another challenge within the geoscience field is that as a percentage of senior management in companies, there are far in between any major oil & gas or independent oil company led by a geoscientist. Primarily geoscientists tend to remain as *highly compensated* subject matter experts, which for many people is a great trade-off.

Information Technology

"Information technology and business are becoming inextricably interwoven. I don't think anybody can talk meaningfully about one without talking about the other" – Bill Gates

Background

Information Technology is a key service and a critical role in the oil & gas industry. Despite having a reputation of being behind in many areas, the oil & gas industry is exciting and ever challenging business that will continue to increase its dependence and adoption of modern technologies.

Information Technology can be subdivided between different groups:

- Commodity services, such as desktop and network support, base applications, email, and IT infrastructure.
- Business or *stream* specific applications, from enterprise software such as SAP or Oracle, to tax systems, billing systems, operational, and logistical systems.
- High Performance computing, particularly in integrated field operations, advanced geological assessments, and petroleum reservoir management.
- Data Analytics, a growing and highly sought-after field where mid to large size companies that collect massive amounts of data in their ERP systems can use that data in a *concise, insightful,* and *value-adding* way.

Educational & Skills Requirements

Educational requirements depend quite heavily on several factors:

- The size of the company, with most large companies requiring at a minimum a bachelor's degree in management or computer information systems.
- Complexity of the IT solution at hand, with more *commodity* type services such as desktop or network support being more open to recruiting people without degrees than say data analytics.
- For specific type of IT products, such as network operations, data security, encryption, or ERP software such as SAP, there are many licenses or courses that are offered by the manufacturers of these products.

- For fields in programming that take many years of practice to master, programming or coding test may be taken before accepting a position to demonstrate a these skills. Commonly sought after programming languages in this industry include, SAP's ABAP, SQL, Java, C++, and even COBOL for programming mainframe applications.

Career Advancement & Stability

Information Technology tends to be one of the most *technically* focused organizations in an oil & gas company and due to the complexity of existing software specific for this industry requires many years of extensive experience.

Most oil & gas companies tend to have a heavy focus on hiring and promoting analysts with a specific set of skills. For example, it is quite difficult for an ABAP programmer to transition from ABAP to say SQL and vice versa. An additional advantage is that careers for widely used applications like SAP ERP software can transition from one sector to the other (i.e. an ABAP programmer's skills are easily migrated from upstream to downstream) or even transition to another industry altogether. Furthermore it can be said that IT careers have one of the most *sector* and *industry* mobility, but one of the lowest *functional* mobility. In other words, an IT professional can go from oil & gas to manufacturing and adapt his or her skills, while it is more difficult to move between skillsets or applications supported or even to other functions within oil & gas.

There are other career paths available that are not entirely based on functional skills but on other more *soft* skills:

- IT project manager, which provide project management services to large IT projects.
- Business IT analyst, which is a hybrid between having business knowledge, having an ability to summarize highly technical material and complexity, and be able to translate it into more widely understood terms.
- Management of Change and training, which depend more on *social skills* than the typical IT position.
- IT management, which is more in line with typical manager roles in other functions that depend on *supervisory* skills, such as ability to see the big picture, plan ahead, desire to develop others, and not rely purely on technical skills. In fact, some of the most successful

IT managers in companies are those that can step back, not get locked up in a technical solution but be able to see a solution on commercial or business value terms.

Typical Job Titles

Information Technology is a wide field with many potential position titles, the ones following are some of the most commonly found in companies:

- IT Analyst
- Financial Systems Analyst
- SAP Programmer
- Systems Engineer
- SAP Functional Analyst

Compensation

The IT function is one of the most highly compensated functions in the oil & gas industry, particularly those that can work with highly complex and critical products such as SAP. Compensation also depends on the type of staff venue an analyst chooses to go through (full time employee with benefits and a *lower salary* vs. contractor, with no benefits and a *higher salary*). Initially most IT professionals start as full time employees since they may not have the necessary experience. After a while they will gain valuable skills in a company, with large companies offering some of the best training programs available. After several years, for highly experienced, qualified and top notch IT professionals, compensation tends to lag and many professionals choose to go through the contract route. If a person has highly sought after skills, is deemed "mission critical" and the application or program supported is in high demand, it is best to pursue a contractors' route. For example, an IT professional's salary might plateau at the low six figures, but as an IT contractor, billable rates can range from $80 to all the way to $400 per hour. Contractors have *higher risks* and no benefits, but many times the high billable rates end up more than compensating for this lack of benefits and high risk.

Business Development & Commercial

> *"Outstanding people have one thing in common: An absolute sense of mission"*- Zig Ziglar

Business Development & Commercial functions are highly critical positions within the following areas of the oil & gas industry:

- In the refined products space, BD & Commercial positions find markets for hydrocarbon products, optimize the value chain and execute supply & trading strategies to ensure that a refinery gets the *highest value* for its products. Conversely, commercial functions also ensure that refineries have the best available crude oil and other feedstocks, at the *best price*, at the *right time*, and at the *right quantities*.
- In the midstream & transportation space, Commercial & BD professionals ensure that hydrocarbons can be moved from one point to another point, at the *best available* transportation rate while providing optionality in terms of having various transportation modes available for capturing the best netback price for hydrocarbons being sold or the lowest price for feedstocks being purchased. From a midstream company perspective, BD personnel ensure that a pipeline has enough *firm capacity* or agreed contracts with shippers so that the pipeline can be built in the first place.
- In the Upstream space, Commercial and BD generally assist with negotiating *Production Sharing Agreements* or PSA for international exploration. For example, BD personnel might help with developing the PSA agreement model to calculate how much the oil company would receive in *cost oil* versus *profit oil*, and at what price they should offer a bonus payment or acquire a lease.
- In the LNG space, Commercial and BD personnel negotiate and manage LNG offtake contracts with LNG buyers around the world so that the LNG facility can be built in the first place. In addition, Commercial personnel ensure that LNG cargoes are scheduled and offtake the LNG on time so that no penalties are assessed on either side. With the growing LNG spot market, trading is also starting to play a much more important role since they can find alternate markets for spot market LNG in other locations.

List of specific functions within Commercial & BD:

- Trading, which is in charge of *buying* and *selling* hydrocarbon commodities in both the physical markets, as well as the financial or paper markets. Trading plays a key role in ensuring any additional barrels not needed or committed with long term contracts for many products are sold or purchased as the best available price possible. Trading is one of the *riskiest* but most *highly* compensated functions within an oil & Gas companies, with trader bonuses sometimes running in the $200,000 to $1-2 million dollars per year depending on performance. In general, oil & gas companies tend to be more risk adverse than purely commercial players such as banks, and thus tend to compensate their traders at a lower rate than financial services. However, in exchange for that, trading positions in the oil & gas industry are far more stable than in the banking industry. In contrast to supply positions, trading tends to have access to inventory of what is generally known as *discretionary barrels*. Discretionary inventory is inventory another function has *not* already earmarked or reserved for contracts. Another function that trading provides is to provide market intelligence to be shared within other groups of supply trading, for example, the trading group might notice that there's a strong seasonal demand for propane in a particular area of the country and share that with the supply group to better optimize contracts.

- Supply functions, ensure that a refinery, marketing operations are supplied with refined products or feedstocks needed to run daily operations and that marketing customers have the product they need. For E&P operations, the supply group administers the company's supply contracts with large crude oil buyers such as refineries, midstream companies, or overseas customers. They maintain these relationships and ensure that volumes from agreed upon contracts can be supplied or purchased without any disruptions to operations. Supply contracts are typically based out of index or *benchmark* prices. Commodity markets for hydrocarbons can be quite different from one another, with different specializations required in crude oil, gasoline, diesel, NGL, chemical feedstocks, sulfur, and natural gas.

- Scheduling, also known as commercial operations, *schedule* and remove any operational barriers for hydrocarbons from flowing from one supply area to the delivery area. Scheduling can work with a variety of modes of transportation, such as wellhead gathering lines, pipelines, trucks, and other marine vessels. Schedulers have to be available 24/7 to assist with any logistical issues that may arise, particularly for pipeline scheduling, since disruptions can occur at any time.

- Market Analysis, the market analysis group provides *timely, accurate,* and *relevant* market information to all groups in the commercial and business development functions. Market analysis groups are usually organized by commodity and region, with a market analysis focusing on a certain NGL component or being responsible for all NGLs in say Conway, KS. The same with other commodities even being subdivided by type of quality, with an analyst being responsible for heavy Canadian crude oils while another for South American crude oils.

- Business Development, business development professionals can work in a variety of projects and sectors. For example, for an E&P company a BD analyst could be responsible for negotiating pipeline transportation rates to gather crude oil from producing fields to the final delivery to a refinery customer. The negotiation process is quite intensive and requires extensive knowledge of the market and when conducting due diligence so that different terms can be negotiated and both the upstream customer, and the midstream company can agree on the terms. Good Excel modeling skills are also needed in order to model long-term revenues and expense streams for a particular deal. Careers within BD can be one of the most rewarding, both in terms of compensation and satisfaction in building new infrastructure.

- Risk is particularly active in the Trading space. Risk ensures that traders do not take unnecessary risk exposure and that an oil company is not unduly exposed to wide fluctuations in the commodities market. Another function that a Risk group performs is to calculate *daily P&L* (profit & loss, also known as mark-to-market) to calculate how much profit or losses a particular trade or trader has brought. Risk functions also monitor compliance with financial and governmental regulations, such as not trading with sanctioned countries or parties and monitoring the trade floor for compliance with *delegation of authority*, making sure that a trade was not executed that was above that individual's delegated approval level. (Say a trade approved a deal worth $1MM, but that trader only has $500,000 authority, then they could get in trouble even all the way to being fired).

Educational & Skills Requirements

Commercial & BD are one of the functions where a *designated* or *specific* degree is not required per se. Professionals in this function can come from a variety of backgrounds, but generally the most widely seen degrees or prior functions are as follows:

- Engineering, particularly chemical engineers, can leverage their skillset in understanding the most technical aspects of the oil & gas industry to engage in commercial activities. Usually a good combination is to have an engineering degree with a Master's in Business Administration.
- Business & Finance degrees, particularly those with good negotiation and market skills are good avenues for getting into these positions.

In addition to a basic bachelor's degree, roles within this function advise having the following skills:

- Good numerical ability, particularly in quickly calculating margins, profits, and other key metrics that lead to success. Trading negotiations are usually done over Instant Messaging and can take from a couple of seconds to a couple of minutes maximum, so having strong *business acumen* and being able to judge quickly whether a deal would be a winner or not is essential.
- Strong emotional intelligence. Nowhere in oil & gas is having this critical skill more essential. A highly skilled individual with low emotional intelligence will not last long in these types of positions. Emotional intelligence involves having the ability to handle emotions very well, handle rejection fairly well, and keep an optimistic outlook as to balance the ups and downs of the business. In addition, an ability to learn quickly from mistakes and not look into perfection is necessary.
- Ability to think outside the box, being independent and persistent is important. Having good judgment, analyzing independent aspects of the matter and not following a "crowd mentality" is also necessary since many traders cannot simply make money by following the crowd.
- Social intelligence is essential in being able to meet customers, foster relationships, being able to reach agreement without unduly jeopardizing an individual's position, but at the same time generating respect is a fine balance.
- Self-confidence also helps tremendously in growing in this field, with outgoing, charismatic individuals having an upper hand over

more introspected and shy individuals. Having a presence and *commanding* the room per se in negotiations whether they are short-term or long-term can mean the difference between an OK career and one that skyrockets.
- Ability to learn quickly and not get bogged down by our individual biases. In trading it does not matter so much whether an individual is ultimately right or wrong, but whether their particular trade or deal made money at a *specific time* and *place*, usually measured in a very short time horizon.

Typical Job Titles

There are many job titles within this function, the titles below are used for illustration purposes:

- Trading or Supply Analyst
- Market Analyst
- Junior Trader
- Senior Trader
- VP of Supply or Trading
- Risk Analyst
- Risk Manager
- Scheduler
- Manager of Scheduling
- VP of Commercial
- Commercial Representative
- Commercial Manager
- VP of Business Development

Career Advancement, Stability & Portability

Careers can advance quite rapidly in the Commercial and BD functions, particularly for those that can demonstrate performance through hard-core P&L measures that can bring in additional revenues or margins. Career advancement within these functions does not typically follow the corporate ladders established for most functions within the industry; instead have their own independent career ladders that reward increasing the perennial bottom line. An additional advantage of most of the career paths available within this function is that measuring performance is quite objective. For example a trader that brings in $4MM in P&L versus another trader that brings in $500,000 per year can be quite easily assessed and ranked, with the first one having a higher compensation structure. In other functions, like accounting or IT, it is a lot more difficult to trace back individual performance to measurable hard results. For this reason, careers within this space can be quite rewarding in both compensation terms and job satisfaction.

A sample career advancement path is shown below for illustration purposes:

- An analyst starts with NGL scheduling to better understand the market, the commodities within NGL, and the particular region or pipeline.
- One year or 18 months later, the scheduler gets promoted to trading analyst, which they work closely with a junior trader in creating trading strategies to capture market dislocations.
- Six months to one year later, that person gets promoted to Junior Trader in helping put financial traders together alongside the senior trader.
- After 1 or 2 years of experience, the junior trader gets promoted to senior trader for NGLs in the Gulf Coast Region.
- After several years, that person might stay in that commodity but compensation will increase substantially as that trader can bring in more and more P&L to the company.
- After several years of experience within the same position, that Senior Trader might decide to move into management within Commercial or stay within that commodity and continue to have a high compensation level.

Another avenue for advancing is to start with Business Development, the below is an example career path:

- A finance graduate joins a Midstream company as an operational analyst to work closely with the BD group.
- After 6-12 months, that analyst gets promoted to Commercial Representative and is responsible for assisting the senior Commercial representative in negotiation existing natural gas gathering & processing contracts for an established producing region.
- After 1-2 years, that individual gets promoted to Manager of Gathering & Processing and is now responsible for leading the

negotiations for natural gas gathering & processing contracts for one of the highest growth producing basins in the country.
- A few years later that Manager gets promoted to another more lucrative commodity, crude oil and leads negotiation and all business development aspects for long-range crude oil pipelines.
- After a decade or so of experience that individual gets promoted to VP of Commercial and is responsible for all commercial and BD aspects of a midstream company.

The previously mentioned career paths are not typical but are certainly possible for *high potential, charismatic, fast learners* with *high emotional intelligence*.

Compensation

As previously discussed, careers within this function can be some of the most highly compensated positions in the oil & gas industry. Salary ranges can be from anywhere $80,000 for basic entry level schedulers to middle 500,000 for some of the most experienced traders and business development personnel. Another factor that influences compensation is the commodity line one specializes in, with higher compensation in commodities where there's a higher market dislocation than others. For example the natural gas trading market used one of the most profitable trading commodities in the United States, with margins or *differentials* between the different producing basins and consuming regions (pre-shale revolution). After the shale revolution, what used to be high differentials for consuming regions disappeared and the business became a lot less profitable.

One of the key advantages of career within this space is the fact that performance is highly objective, unlike most other careers within the oil & gas industry. For example, a trader may not get along with other co-workers, be rude, impolite, arrive late at work and so forth, but as long as that individual can bring in substantial income to the company that individual's behavior will be overlooked or dismissed. The same cannot be said for other positions in say accounting, HR, IT, or even engineering. Moreover, other aspects of evaluation such as career history (i.e. a job hopper trader) are not a significant factor since these individuals can start contributing to the *bottom line* right from *the start*.

Chapter VIII – Conclusion

"Inaction breeds doubt and fear. Action breeds confidence and courage. If you want to conquer fear, do not sit home and think about it. Go out and get busy"
– Dale Carnegie

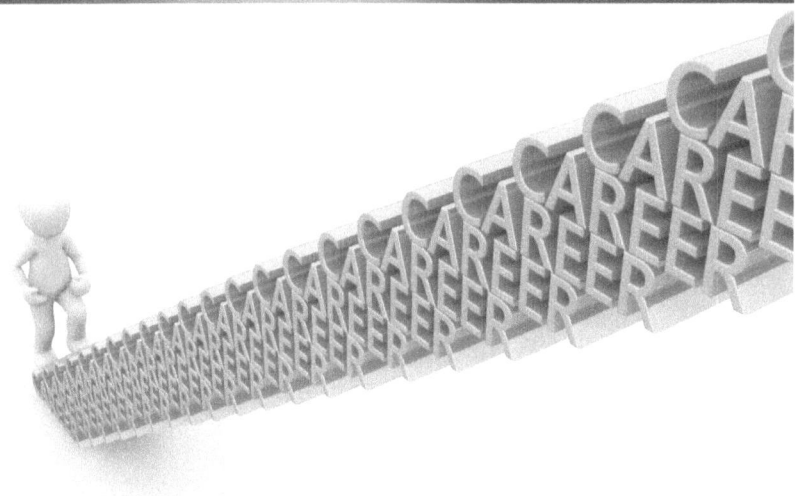

Summary

A career in the Oil & Gas industry can be quite rewarding whether you work in upstream, midstream, downstream, or service sectors. As we saw in the prior chapter choosing the right function can mean quite a difference in long-term compensation and job satisfaction. We also saw how employee rankings and ratings are critical for an employee's long-term compensation and promotion progression.

Assessing your current & future career state

There are several questions that you can ask yourself when assessing the current state of your career:

- Reviewing your past positions and see how much you have grown in comparison to how you started. Have you gained valuable skills that increase your marketability for future positions within your company or other companies?
- At what stage in your life are you? Are you having kids so have a higher need for work life balance? Does your current position provide for excellent work life balance? If so, you may consider keeping your existing position since the possible increase salary may not compensate for the higher hours required.
- Are you considered a go-to person in your current person? Do you get satisfaction from being able to respond to questions or challenges at work faster than everybody else? This is something you need to factor in your assessment of current state. If you move to a new position, would you be able to handle for the first few months (or even years, depending on the level of complexity) to not be seen as the *go-to person* when compared to your prior position? For many people transition into an entirely new year, this lack of validation for the first few months or years can take a toll on individuals that a have a high external validation dependency.
- Where do you see yourself in 5 years? Or in 10 years? Does your current provide a basic building block for you to gain the skills that will be used in the future?
- If you could only have one job for the rest of your life, would you be OK with doing the same tasks that you do today?

Questions to ask when selecting a company to work for

As repeated throughout this book, choosing the right company from the beginning of your career can have great impacts long term in your overall career, as well as your personal life. Here are sample questions to ask:

- How does Executive Management get promoted? Do they promote from within or do they have a lot of external hires?
- How much emphasis does the company place on new hires vs. mid-level career employees?
- For your particular recruiter, assuming he or she works in the same function, ask them about their career, how many locations have they been?
- What is the likelihood of receiving a *cross-functional* assignment?
- What are the typical first assignments or rotations in this company or department? If the answer seems like a lot of data gathering, or repetitive task-based assignments, you may want to re-assess the skills you would get by working there.
- How many meetings do employees frequently participate in?
- Can you tell me about some of the skills you have gained in your company? What new systems, processes, or businesses have you learned during your stay with the company?
- What do you wish you could change in your company?
- Which cities or countries does this company have operations?
- How often and at what stage in their career do people in my function transfer to these locations?

Behaviors to observe when selecting a company to work for

Asking questions would usually receive a standard PR/HR response about career development, but it is also important to *observe* how employees behave and what behaviors they exhibit:

- What is the attitude and behaviors of mid-level career employees (Those with 5-7+ years with the company)?. Do they seem happy and fulfilled?
- Does the company provide unrealistic & advancement opportunities for new hires that do not correspond with reality? I.e. being a manager within 2-3 years?

- Do new hires seem to have less optimism as the *honeymoon* period wanes off or do they keep the level of optimism?
- When interacting with recruiters, do you get emails at odd hours? If so, this may mean lack of a work life balance in that company.
- Do employees seem to be willing to go the *extra mile*, no matter how small or big the request is? If not, this might mean disgruntled employees which translate into low job satisfaction.
- Look at the most coveted or sought-after positions within your function or department. How long does it take to get there? Who gets promoted there?
- Does the company seem to value *perception* over *reality*? Who gets promoted will provide insights on what the company looks for.
- How reliant is the company on meetings?
- Does the company place an unduly high level of attention on new hires and interns? If so, this should be evaluated with caution and compare to the level of interest the company places on mid-career employees. Pretty soon you will be a mid-career employee, and if the company's strategy is to forget about these employees you may want to reevaluate your decision.
- Do they make certain people wait for a career rotation while others advance faster? Then you might not be considered a high potential employee.
- Is management prone to *blindly* follow general fads or following the latest and coolest trend versus being stubborn, or independently minded?

Questions to ask a particular department

- How many people transfer in and out of this particular department?
- How many SME's does this department have and how much do they rely on these people?
- What is the expected career path for most employees in this department?
- How difficult is it to achieve proficiency in a particular area compared to others?
- Who does the department report to and, in turn, what department reports to us?
- How long has the current supervisor been here?

- Will this position be the last of the new hires?
- Why is this particular position being filled?

The Importance of an Action Plan
Every person who would like to have a course of action for their career should start developing an action plan at any stage in their career, from finishing high school to changing careers in mid or senior level careers.

Action Plan for Early Career
Here are listed several good recommendations for early career employees:

- Focus on acquiring skills, knowledge, and understanding of the business. For the first few months most new employees' contributions are modest when compared to the investment that the company is putting into them.
- No task is low or "boring" enough for an early career employee. As the late Steve Jobs used to say "You can look back to connect the dots, you cannot look forward". Every single task or assignment, not matter how mundane or basic it might seem, allows you to gain knowledge on the company's operations, the industry, and how the overall job flows entirely.
- Spend extra hours at the beginning of any new position to accelerate learning. If you can learn, absorb, and apply new knowledge faster than your peers, you will be ahead of your peers.
- Take notes and conduct extra research (on your own time) about your group, function, organization chart, systems, and processes. You can make very good first impressions the sooner you increase your proficiency in your role.
- Have patience as you learn new things and start to apply them. For many high achievers that like to achieve many goals at the same time.
- Do not be afraid to ask questions, not only as you start but as you gain experience you will be able to refine questions and learn more.
- Be polite when proposing new ideas or challenging existing ideas. As you will be seen as the least experienced staff member in your group, you need to be conscious of how you will be perceived.
- Have patience for the first few years of your career as you gain new skills.

Action Plan for Mid-Level & Late Career

Mid-level career employees have some of the biggest challenges involved in developing their careers and there are many reasons for that:

- Mid-level career people are thought as being too experienced to switch careers or change career paths, but less experienced than those considered as subject matter experts.
- Late career people are thought as being resistant to change, so demonstrating an ability to execute change management is flexible, particularly in those functions that rely on new technology.
- If you like technology in particular, it is important to keep up with changes and brand yourself appropriately. Many recruiters may have a bias or misconception of more experienced employees as being with outdated skills.
- Particularly in modern society, people are attracted to *novelty* and the *new* versus what is *dependable* and *predictable*, so this is a barrier that mid-level career and late career employees have to face and overcome. Being familiar or proficient with the latest technologies or topics within your specific function or sector can help you brand yourself as modern and in-tune with change.
- Being open to change methods and conveying an ability to learn new skills can make a significant difference in how you are perceived when recruiting or being interviewed.
- Use your dependability as a marketing device in comparison to new workers which are seen as less reliable. Create rapport with recruiters of the same generation or similar interests, so that you can be seen as the more stable candidate.
- Use your experience and, mention prior transformational changes that you have experienced in your career that may not be available for younger workers. For example, if you have been through a merger, acquisition or disposition be sure to use relatable examples of how those experiences could help you in a future position or future company.
- You should have a strong network, and think of ways that could help your new employer.
- Talk to younger professionals in your same career path to see what employers are looking for when they hire them. Certifications, degrees, classes, a certain type of experience.

- Make it clear that you are here to prove yourself. You aren't looking to get hired for what you have done; rather you have so much left to offer.
- Offer to mentor younger employees.
- Offer to build a database with your various experiences. Even if it is not oil and gas related, many things will still apply.
- Be willing to listen to the younger crowd.
- Embrace learning new technology.
- You might be older than some or all of your bosses. Make them feel at home and give them the same respect that you would want.
- Talk about the future of the company that will long exceed your time there. For instance, "I am so excited to see where this company will be in 50 years." Yes, everyone knows that a 65-year old probably will not be around for that. However, it shows that you are not looking towards retirement and are still interested in the long term success of the company.

Last Words

Ryan and I wish you all the best in your future job searches, whether that is within your existing company or with another company. Regardless of the stage or condition of your career, it is always important to keep optimism up and to not get bogged down with pessimism. A career in the oil & gas industry can be one of the most rewarding and satisfying careers of any industry. Keep up the good work with your career and we look forward to hearing you from you. Please do write us at book@oilgascareers.net.

Alfonso & Ryan

June 2018

Glossary

401K Plan: An employer-sponsored retirement plan that has become an expected benefit and is therefore important in attracting and retaining employees. A 401(k) plan allows employees to defer taxes as they save for retirement by placing before-tax dollars directly into an investment account. Employers also contribute to the plan tax-free, for instance by matching contributions. Some plans enable employees to direct their own investments. These plans can be expensive and complex to manage. It is common for companies to outsource all or part of their plan[177].

Applicant Tracking System (ATS): A software application that began as a way to electronically handle recruitment needs but has since expanded to the entire employment life cycle. Onboarding, training and succession planning capabilities now exist, for example. An ATS can be implemented on an enterprise level or small business level, depending on the size and needs of the company[178].

Attrition: A gradual voluntary reduction of employees (through resignation and retirement) who are not then replaced, decreasing the size of the workforce[179].

Barrel/BBL: Volume unit corresponding to 42 U.S. gallons or 159 liters. A U.S. barrel is widely used in the industry

Behavioral-Based Interview: An interview technique used to determine whether a candidate is qualified for a position based on their past behavior. The interviewer asks the candidate for specific examples from past work experience when certain behaviors were exhibited[180].

Behavioral Competency: The behavior qualities and character traits of a person. These act as markers that can predict how successful a person will be at the position he/she is applying for. Employers should determine in advance what behavioral competencies fit the position and create interview questions to find out if the candidate possesses them.

---

[177] http://www.hrmarketer.com/glossary-of-hr-and-benefits-terms/
[178] Ibid
[179] Ibid
[180] Ibid

Benchmark Job: A job commonly found in the workforce for which pay and other relevant data are readily available. Benchmark jobs are used to make pay comparisons and job evaluations[181].

Biodiesel: A renewable fuel produced from vegetable oils or animal fats that can be blended with petroleum-derived diesel to produce biodiesel blends for use in diesel engines. Pure biodiesel is referred to as B100, whereas blends of biodiesel are referenced by how much biodiesel is in the blend (e.g., a B5 blend contains five volume percent biodiesel and 95 volume percent ULSD).

BOE Barrel of Oil Equivalent: BOE is used as a standard unit to measure combined oil and natural gas. The latter is converted from standard cubic meters into barrels of oil equivalent using a certain coefficient

BPD: Barrels Per Day

Brent Crude Oil: a light, sweet crude oil, though not as light as WTI. Brent is the leading global price benchmark for Atlantic basin crude oils.

BTU: British Thermal Unit. A BTU is equivalent to the energy required to heat 1 pound of water by one degree Fahrenheit. A BTU can also be defined as the amount of energy released by striking a single wooden match.

Condition of Employment: An organization's policies and work rules that employees are expected to abide by in order to remain continuously employed.

Confidentiality Agreement: An agreement between an employer and employee in which the employee may not disclose proprietary or confidential information.

Condensates: Condensate is a mixture of hydrocarbons that exists in the gaseous phase at original reservoir temperature and pressure, but when produced, is in the liquid phase at surface pressure and temperature.

Core competencies: The particular set of strengths, experience, knowledge and abilities that differentiate a company from its competitors and provide competitive advantage. Employees should possess these qualities in order to advance business goals.

---

[181] Ibid

Crude Oil: Oil is a mixture of molecules of hydrogen and carbon that are primarily found in liquid state at atmospheric conditions. The hydrocarbon mixtures found in crude oil range in properties, such as boiling points and number of carbon and hydrogen molecules.

Cubic Foot: a unit of measure for volume, usually used to measure natural gas. It is the space or volume a cube of 1 by 1 by 1 occupies or in other words 1 foot long by 1 foot wide by 1 foot in height (LxWxH). Since a cubic foot is a very small amount of gas volume, gas is more commonly measured in terms of one thousand cubic feet or 1MCF

Cubic Meter: An international metric system unit of measure for volume, commonly used for measuring natural gas. One cubic meter is equivalent to 35.31 cubic feet

DAO: De-Asphalted Oil

DCS: Distributed Control Systems, system in the refinery or other plant that controls processes in a facility.

Deferred compensation: Payment for services under any employer-sponsored plan or arrangement that allows an employee (for tax-related purposes) to defer income to the future.

Defined Benefit Plan: A retirement plan that pays participants a lump-sum amount that has been calculated using formulas that can include age, earnings and length of service.

Defined Contribution: A pension plan that clearly defines the amount of contributions, which is usually a percentage of an employees' salary. The benefits payable at retirement depend on several factors including future investment return and annuity rate at retirement.

Downstream: Usually refers to refining and marketing of hydrocarbons as well as the production and marketing of petrochemicals.

EIA: United States Energy Information Agency

EPA: Environmental Protection Agency. EPA regulations impact many areas in the oil & gas industry, particularly in setting renewable standard requirements, gasoline composition and other areas.

Exempt Versus Non-Exempt Employees: The difference between exempt and nonexempt employees is who gets paid overtime and who doesn't. The

U.S. Department of Labor specifically designates certain classes of workers as exempt, including executives, administrative personnel, outside salespeople, highly skilled computer-related employees, doctors, lawyers, engineers, etc. Managers who hire and fire employees and who spend less than half their time performing the same duties as their employees are typically also exempt employees. In general, the more responsibility and independence or discretion an employee has, the more likely the employee is to be considered exempt. Generally, any worker performing repetitive tasks is most likely nonexempt and must be paid overtime.

Exit Interview: The final meeting between management, usually someone in the HR department, and an employee leaving the company. Information on why the employee is leaving is gathered to gain insight into work conditions and possible changes or solutions.

Expatriate: An employee who is transferred to work abroad on a long-term job assignment.

Exploration of oil and natural gas: Exploration that includes land surveys, geological and geophysical studies, seismic data gathering and analysis and well drilling.

Feedstocks: Crude oil and petroleum products used as inputs in refining and petrochemical processes.

FERC: The Federal Energy Regulatory Commission.

GAAP: Generally Accepted Accounting Principles. Usually meant to refer to either United States Generally Accepted Accounting Principles (U.S. GAAP) or International Financial Reporting Standards (IFRS). Companies can usually disclose both GAAP and non-GAAP measures depending on the audience and document that is used for.

Jobbers: Retail stations owned by third parties that sell products purchased from or through us.

Light/Medium/Heavy Crude Oil: Terms used to describe the relative densities of crude oil, normally represented by their API gravities. Light crude oils (those having relatively high API gravities) may be refined into a greater amount of valuable products and are often more expensive than a heavier crude oil.

LNG: Liquefied Natural Gas is obtained through the cooling of natural gas to minus 260 degrees F at normal pressure. The gas is liquefied to allow transportation from the place of extraction to the sites at which it is transformed back into its natural gaseous state (re-gasified) and then used by consumers for different purposes.

LPG: Liquefied Petroleum Gas usually meant to describe a combination of propane and butanes. LPG is a key source of cooking fuel around the world where there is no natural gas (methane gas) pipeline distribution.

Margin: The difference between the average selling price and direct acquisition cost of a finished product or raw material excluding other production costs (e.g. refining margin, margin on distribution of natural gas and petroleum products or margin of petrochemical products). Margin trends reflect the trading environment and are, to a certain extent, a gauge of industry profitability.

MCF: One thousand cubic feet. M stands for Roman numeral M or mille for one thousand.

Metric Ton: A measure of mass or weight. A metric ton in this book is converted using a 7.33 barrel per metric ton conversion. The conversion from metric ton to barrel has to factor in the density of the referenced liquid.

Midstream: Usually refers to intermediate processing and transportation of oil & gas products, whether in raw state, intermediate or finished state.

MMBOED: millions of barrels of oil-equivalent per day. Natural gas is converted to a barrel-oil equivalent using a 6,000 cubic feet per barrel conversion factor.

MMBTU: Million BTU. One M stands for Roman numeral M; therefore two Ms stand for one million.

MSCF/d: Abbreviation for a thousand standard cubic feet per day, a common measure for volume of gas.

Natural gas liquids (NGL): Liquid or liquefied hydrocarbons recovered from natural gas through separation equipment or natural gas treatment plants. Ethane, propane, normal-butane and isobutane, isopentane and pentane plus are natural gas liquids. NGLs can also be refined from crude oil at a refinery.

Natural Gas: a mixture of hydrocarbons, primarily found in gaseous form. Natural gas in raw form is primarily composed of methane gas, but can also contain ethane, propane, iso-butane, normal butane and pentanes and heavier molecules. When most people refer to natural gas, they are usually referring to "pipeline quality" gas, which is primarily methane and a little bit of ethane, since most of the heavier molecules would have to be removed to prevent "slugs" of liquids clogging and compromising a pipeline's safe operations.

New York Mercantile Exchange (NYMEX): A major commodities futures exchange, where contracts for commodities such as crude oil, natural gas and refined products are bought and sold every day.

Observation interview: A method of assessing job requirements and skills by observing the employee at work, followed by an interview with the employee for further assessment and insight.

Offshoring: The act of moving work to an overseas location to take advantage of lower labor costs. Offshoring usually involves manufacturing; information technology and back-office services like call centers and bill processing. Companies can build its own work center abroad, establish a foreign division, or create a subsidiary in remote locations.

Onboarding: The process of moving a new hire from applicant to employee status ensuring that paperwork is done, benefits administration is underway, and orientation is completed.

Outsourcing: Contracting out non-core functions, such as payroll, benefits administration or manufacturing, to save money and focus on what the company does best.

RVP: Reid Vapor Pressure

SEC: United States Securities & Exchange Commission is an independent agency of the U.S. Federal Government that oversees securities exchanges and enforces U.S. federal securities laws. The SEC regulates company issued documents, such as form 10-Ks, 20-Fs, annual reports, earnings releases and other externally published documents.

Straight run: product produced off of the crude or vacuum unit and not further processed.

Sweet/Sour crude oil: Terms used to describe the relative sulfur content of crude oil. Sweet crude oil is relatively low in sulfur content; sour crude oil is relatively high in sulfur content. Sweet crude oil requires less processing to remove sulfur and is usually more expensive than sour crude oil.

Take-or-pay: Clause included in natural gas supply contracts according to which the purchaser is bound to pay the contractual price or a fraction of such price for a minimum quantity of gas set in the contract whether or not the gas is collected by the purchaser. The purchaser has the option of collecting the gas paid for and not delivered at a price equal to the residual fraction of the price set in the contract in subsequent contract years.

Talent Management: Also called Human Capital Management, the process of recruiting, managing, assessing, developing and maintaining employees.

Team building: A philosophy of job design which fosters teamwork to create a work culture that values collaboration. It is a training program designed to encourage employees to view themselves as members of interdependent teams instead of as individual workers, in which people understand and believe that thinking, planning, decisions and actions are better when done cooperatively.

TAN: Total Acid Number

Throughput: The quantity of crude oil and feedstocks processed through a refinery or a refinery unit.

Turnaround: A periodic shutdown of refinery process units to perform routine maintenance to restore the operation of the equipment to its former level of performance. Turnaround activities normally include cleaning, inspection, refurbishment, and repair and replacement of equipment and piping. It is also common to use turnaround periods to change catalysts or to implement capital project improvements.

U.S. Gulf Coast Pipeline CBOB: A grade of gasoline blendstock that must be blended with 10% biofuels in order to be marketed as Regular Unleaded at retail locations.

U.S. Gulf Coast Pipeline No. 2 Heating Oil: A petroleum distillate that can be used as either a diesel fuel or a fuel oil. This is the standard by which other Gulf Coast distillate products (such as ultra-low sulfur diesel) are priced.

Ultra-Low Sulfur Diesel (ULSD): Diesel fuel produced with a lower sulfur content (15 ppm) to reduce sulfur dioxide emissions. ULSD is the only diesel fuel that may be used for on-road and most other applications in the U.S.

Upstream/Downstream: The term upstream refers to all hydrocarbon exploration and production activities. The term downstream includes all activities inherent to the oil and gas sector that are downstream of exploration and production activities.

Work-life Balance: The attempt to balance work and personal life in order to have a better quality of life. A person with a balanced life is an asset to his or her business, as he or she experiences greater fulfillment at work and at home.

Work/Life Employee Benefits: Work/Life benefits are "non-traditional" employee benefits that assist employees in managing their lives. Employers purchase these services from vendors and they are offered to employees as benefits. These services can make the difference in attracting and retaining employees. Common life management benefits include: child and elder care referral services, employee assistance program (EAP), concierge, legal assistance, and emergency back-up childcare.

Workforce Planning: The assessment of the current workforce in order to predict future needs. This can consist of both demand planning and supply planning. Many e-recruitment software providers include modules for workforce planning.

Wrongful Termination: A legal term referring to when an employee was fired for an illegal reason.

West Texas Intermediate Crude Oil (WTI): A light, sweet crude oil characterized by an API gravity between 38 and 44 and a sulfur content of less than 0.4 weight percent that is used as a benchmark for other crude oils.

www.ingramcontent.com/pod-product-compliance
Lightning Source LLC
Chambersburg PA
CBHW031621210526
45464CB00004B/1682